IUTAM Symposium on Field Analyses for Determination of Material Parameters – Experimental and Numerical Aspects

SOLID MECHANICS AND ITS APPLICATIONS
Volume 109

Series Editor: G.M.L. GLADWELL
Department of Civil Engineering
University of Waterloo
Waterloo, Ontario, Canada N2L 3GI

Aims and Scope of the Series

The fundamental questions arising in mechanics are: *Why?*, *How?*, and *How much?* The aim of this series is to provide lucid accounts written by authoritative researchers giving vision and insight in answering these questions on the subject of mechanics as it relates to solids.

The scope of the series covers the entire spectrum of solid mechanics. Thus it includes the foundation of mechanics; variational formulations; computational mechanics; statics, kinematics and dynamics of rigid and elastic bodies: vibrations of solids and structures; dynamical systems and chaos; the theories of elasticity, plasticity and viscoelasticity; composite materials; rods, beams, shells and membranes; structural control and stability; soils, rocks and geomechanics; fracture; tribology; experimental mechanics; biomechanics and machine design.

The median level of presentation is the first year graduate student. Some texts are monographs defining the current state of the field; others are accessible to final year undergraduates; but essentially the emphasis is on readability and clarity.

For a list of related mechanics titles, see final pages.

IUTAM Symposium on

Field Analyses for Determination of Material Parameters – Experimental and Numerical Aspects

Proceedings of the IUTAM Symposium held in
Abisko National Park, Kiruna, Sweden,
July 31 – August 4, 2000

Edited by

P. STÅHLE

Malmö University, Malmö, Sweden

and

K.G. SUNDIN

Luleå University of Technology, Luleå, Sweden

KLUWER ACADEMIC PUBLISHERS
DORDRECHT / BOSTON / LONDON

A C.I.P. Catalogue record for this book is available from the Library of Congress.

ISBN 1-4020-1283-7

Published by Kluwer Academic Publishers,
P.O. Box 17, 3300 AA Dordrecht, The Netherlands.

Sold and distributed in North, Central and South America
by Kluwer Academic Publishers,
101 Philip Drive, Norwell, MA 02061, U.S.A.

In all other countries, sold and distributed
by Kluwer Academic Publishers,
P.O. Box 322, 3300 AH Dordrecht, The Netherlands.

Printed on acid-free paper

Printed in the Netherlands.

Contents

Experimental Methods

Optimization Techniques and Simulation Methods

Material Modeling

Executive Committee

P. Ståhle, Chairman, Malmö University, Sweden
K.G. Sundin, Luleå University of Technology
P. Isaksson, Malmö University, Sweden
J. Kajberg, Luleå University of Technology

Organising Committee

P. Ståhle, Chairman, Malmö University, Sweden
S. Aoki, Tokyo Institute of Technology, Japan
H. D. Bui , Ecole Polytechnique, France
J. Engelbrecht, Institute of Cybernetics, Estonia
J. Field, Cavendish Labs, Cambidge, England
J.F. Kalthoff, University of Bochum, Germany
K.-S. Kim, Brown University, USA
N. Morozov, St Petergsburg State University, Russia
M. Ortiz, California Institute of Technology, USA

PREFACE

The symposium was motivated by the increasing need for modelling of material behaviour under various mechanical conditions. This need is driven by the evolution of computer capacity and the resulting ability for engineers and scientists to address complex problems. Reliable models for material behaviour, including accurate numerical values of parameters, are necessary for a continued beneficial development of the computational side of solid mechanics. High rate plasticity, thermally assisted creep and phase transformations are only a few examples of areas where more accurate models are needed.

Experiments are necessary for the establishment of models and parameters, and modified versions of conventional test methods can make important contributions. Also modern optical methods offer a high potential for future experimental development. Numerical simulations of experiments and so-called inverse modelling are also frequently used techniques.

The aim of the symposium was to bring together researchers with an interest in the area of experimental and computational aspects of material modelling for exchange and discussion of promising methods and results.

Abisko, a national park in the Swedish mountain district about 200 km north of the arctic circle and about one hour drive from the airport of Kiruna, was chosen for the symposium. The tourist hotel in the park, overlooking a beautiful lake, offered a suitable venue for the symposium. This environment with tracks for short walks (and long hikes), goals for small excursions and a hotel with restaurant and bar gave the delegates many opportunities to meet, socialise and discuss during breaks and evenings.

There were 30 delegates, of which 19 were authors; 20 papers by a total of 45 authors were presented at the symposium. The papers covered a wide scientific field within the scope of the symposium, and many different aspects of material modelling were illuminated. Extensive discussions after each presentation were encouraged.

In the final discussion on the last day it was concluded that this research field is an important one for the future development of solid mechanics, both in industrial applications and in scientific research. The importance of understanding the physical phenomena governing material behaviour was of course recognised, but empirically based modelling in more or less narrow application areas is motivated and of practical importance if physical understanding is lacking. Such models should be as mathematically simple as possible to facilitate easier determination of parameters. It was also stressed that a better communication between researchers in experimental mechanics, computational mechanics, solid mechanics and material science is desired for the development of the area of material modelling.

Financial support from Volvo Aero Corporation Sweden and The Knowledge Foundation is gratefully acknowledged.

Per Ståhle and K G Sundin

WHOLE-FIELD DISPLACEMENT MEASUREMENT WITH MOIRÉ INTERFEROMETRY AS A BASIS FOR A MICROSTRUCTURAL FE MODEL OF A POLYMER BONDED EXPLOSIVE

H.T. Goldrein, P.J. Rae, S.J.P. Palmer
Cavendish Laboratory, Madingley Road, Cambridge CB3 0HE, UK

A.L. Lewis, G. Miles
AWE Aldermaston, Reading, Berkshire RG7 4PR, UK.

N. Zahlan
ICI Technology, PO Box 90, Wilton, Middlesborough, Cleveland TS90 8JE, UK

1. INTRODUCTION

Polymer bonded explosives (PBXs) are highly filled composite materials of crystalline high explosive in a polymeric binder matrix. The polymer binder provides structural integrity and mechanical strength. Although the polymer content of a PBX may be as low as 5 to 10 volume percent, the chemical and physical structure of the polymer binder, and the nature of its interaction to the crystalline explosive filler, dominate the mechanical characteristics of the composite. PBXs find important application as the main charge explosive in some weapons, and may be required to fulfil an engineering role, as well as meeting other stringent requirements.

On the micron-scale, PBXs typically have a highly heterogeneous structure, embodying a range of explosive filler particle shapes and sizes, with consequent variations in the thickness and nature of the inter-particle binder layers. For these reasons the bulk mechanical properties can be hard to predict with precision. Interpretation and comparisons of test data from standard mechanical tests can be further complicated by the effects of ageing, changes to manufacturing processes or source materials, or even pre-test damage arising from stresses imposed on these relatively brittle materials during specimen preparation.

Investigation of PBXs at the microstructural level by both experiments and modelling has provided key insight into the influential relationships between details of the molecular and microstructure of the PBX constituents, and the macroscopic properties of the bulk material. In this approach, high-resolution moiré interferometry has been used to measure whole-field displacements with a sensitivity of order of tens

P. Ståhle and K.G. Sundin (eds.), IUTAM Symposium on Field Analyses for Determination of Material Parameters - Experimental and Numerical Aspects, 1–10.

of nanometres and a spatial resolution of approximately a micron. With the sponsorship and guidance of AWE Aldermaston, innovative experiments at Cambridge University to visualise the details of PBX microstructure and measure localised deformations under quasistatic loading have been modelled at ICI Technology with significant success. The development of a finite element model for the microstructure offers a powerful tool for unravelling the intricacies of microstructural effects: using the model, it is possible to isolate and examine the influence of particular microstructural features (such as binder interlayer thickness, or explosive crystal size) on predicted deformation characteristics. The experimental data from Cambridge is used to improve the accuracy of and validate the model. We report here the experimental technique, the preliminary modelling studies, and some early observations of microstructural effects. Both the experimental methods and the modelling have broad applicability to microstructural studies of a variety of composites.

2. THE BRAZILIAN TEST

Due to the inherent danger of testing explosives, experiments at the Cavendish are limited to samples of less than approximately 0.5 g, making the testing of uniaxial specimens impractical. The "Brazilian" or "diametral compression" test is used instead. It is a biaxial method for estimating the tensile failure stress of a material by applying a compressive load diametrically to a cylindrical sample. A vertical tensile failure is generated in brittle materials as shown in Figure 1.

Simple elasticity theory[1] applied to plane anvils predicts that the tensile stress at the centre is a maximum given by

$$\sigma_x = \frac{2P}{\pi D t} , \tag{1}$$

where P is the applied force, D is the sample diameter and t is the sample thickness. The compressive stress at this point is three times the tensile one. It has been shown[2] that curved anvils tend to lower the shear stresses at the anvils, which in turn prevents premature sample failure. The effect on the central stress state has been investigated in the same reference and the following expression derived

$$\sigma_x \approx \frac{2P}{\pi D t}\left\{1-\left(\frac{b}{R}\right)^2\right\}, \tag{2}$$

where b is the contact half-width of the anvils and R is the radius of the specimen. This b/R ratio must be measured for the material being tested. The curvature of the anvils is chosen so that the fraction is greater than 0.27 since this is required for the failure to be purely tensile.

In all of the experiments a b/R ratio of 0.3 is assumed based on past tests with PBXs. Substituting this value into (2) gives a correction factor of 0.91 on the stress predicted by equation (1). The tests described here have been on samples nominally 10 mm in diameter and 4 mm thick. The samples were of a highly-filled PBX based on the crystalline explosive HMX (cyclotetramethylene-tetranitramine) with a rubbery polyurethane binder.

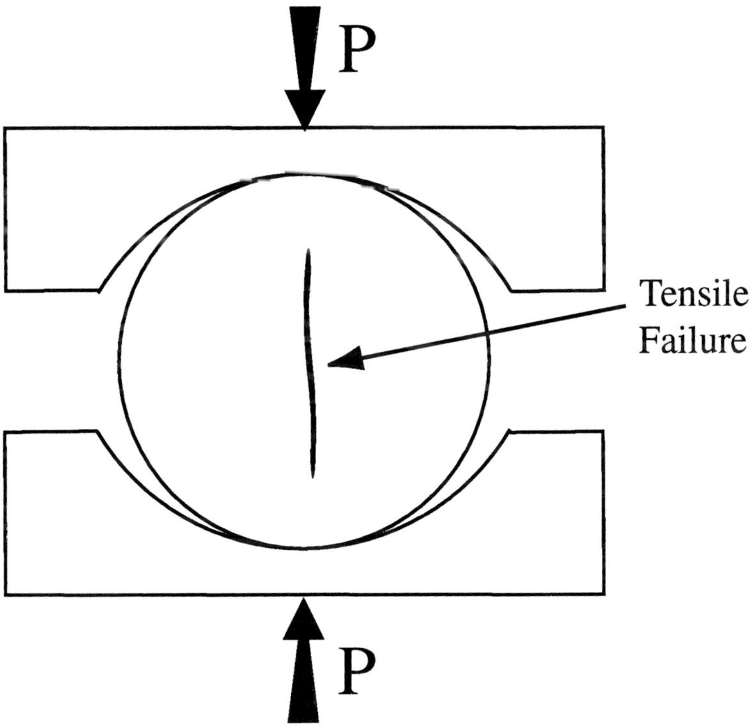

Figure 1. Brazilian test geometry

Previous work comparing the low strain rate tensile strengths of PBXs using the conventional dumbbell uniaxial geometry and the Brazilian test has shown a good correlation over a range of materials[3]. It appears that the tensile strengths obtained using the uniaxial geometry are consistently a factor of approximately two times higher than those obtained with the Brazilian test. Within the constraints imposed by testing hazardous materials in the laboratory, this method provides valuable comparative data from small samples. The work presented here continues from previous studies on PBXs carried out in the Cavendish Laboratory[4-7].

3. MOIRE INTERFEROMETRY

High-resolution moiré interferometry is a sensitive coherent optical technique, which allows the measurement of in-plane displacements[8-10]. The sample is imaged through a low power microscope giving a spatial resolution of approximately 1 μm, with displacement sensitivity approaching 10 nm. This arrangement allows a direct correlation of features in the composite microstructure with the measured displacement field.

Figure 2 shows a diagram of a simple moiré interferometer. The technique works by interrogating a high frequency phase grating cast onto the specimen surface using two collimated beams derived from a single laser source. The +1 diffracted beam from one incident beam and the –1 beam from the other are set up such that the beams leave normal to the grating surface. Any change in the grating pitch, produced by mechanical strain for instance, will cause a change in the angle at which the beams leave the grating. The resulting fringe pattern produced by the interference between the beams is proportional to the deformation of the grating[8]. This fringe pattern may be recorded onto film or with a CCD camera. Each fringe represents a local displacement of half a grating pitch, in a direction perpendicular to the grating rulings, so a 1200 lines mm^{-1} grating will give a sensitivity of 0.417 μm per fringe. The system is insensitive to rigid body motion of the specimen, since the grating pitch does not change, and to out of plane motion, since the path length of each laser beam is equally affected leading to an unchanged interference pattern. Using computer analysis and phase stepping, a sensitivity of around one hundredth of a fringe may be achieved, leading to a displacement sensitivity approaching 10 nm. Our optical arrangement allows one component of in-plane displacement to be measured. To measure both components of displacement more complicated optics are required.

The gratings are cast onto polished and stained PBX samples by replication from a 1200 lines mm^{-1} linear silicone rubber sub-master. PL-3 epoxy (Measurements Group UK, Basingstoke, Berks.) was used since it is highly compliant (modulus 1.4 MPa) and of low viscosity when uncured. In this way thin gratings may be replicated (<5 μm) which do not reinforce the specimen surface or cause the local strains to become smeared out due to shear lag errors. The gratings are coated with a thin layer of gold (3–4 nm) to enhance the diffraction efficiency, whilst still allowing white light pictures to be obtained through the grating.

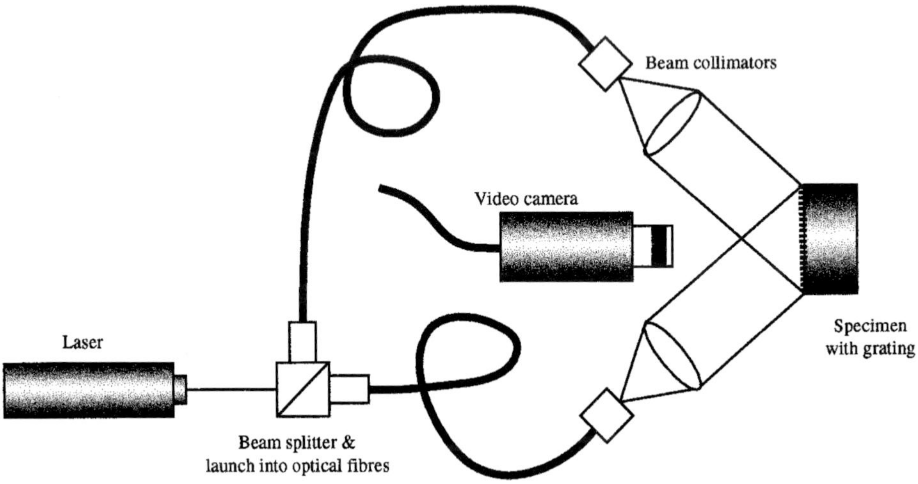

Figure 2. Schematic of a simple moire interferometer

Local strains may be related to the underlying microstructure by obtaining white light micrographs at the same time as the laser interferograms from which moiré measurements are made. See

Figure *3* for a typical high-magnification displacement map from a loading sequence. The majority of the strain is concentrated in what appears to be fine-grained material. Significant concentrations of strain can be seen around the boundaries of the larger filler particles, and in addition, numerical analysis shows that the finer grained regions support much larger strains than the bulk material.

The kinks in the displacement contours across the central crystal are evidence of twinning and of significant strain in the relatively stiff crystalline explosive. The twinning is shown clearly post-failure, using reflection Nomarski interference, as illustrated in Figure 5. Earlier research on twinning is discussed in Ref. 11.With this series of material results, it was possible to construct a finite element (FE) model of the microstructure of the specimen to model the deformation and to study the sensitivity of the model to various material parameters.

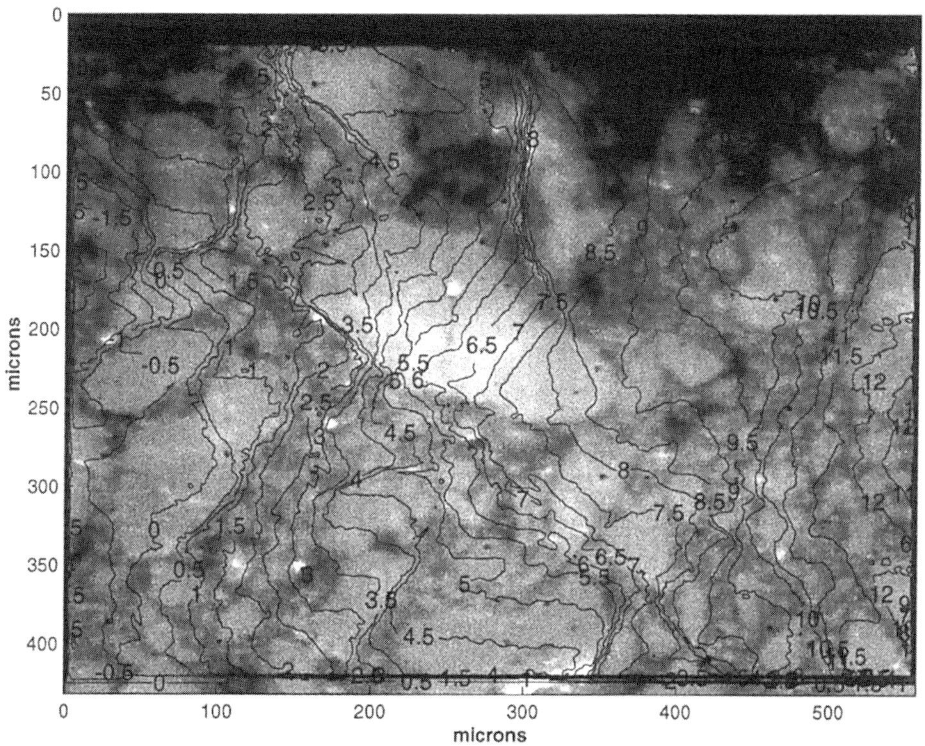

Figure 3: Displacement map for sample at a load of 48.9N. contour interval = 0.5 μm

4. FINITE ELEMENT MODELLING

A major challenge to the modelling of this experiment is the complexity of the Brazilian test. Whilst equation (2) above gives an analytical expression for the stress at the centre of the specimen, it is not sufficient to assume that the stress is constant across the whole region of interest being studied. Consequently, it was first necessary to construct a bulk model of the Brazilian test to determine the boundary conditions to be applied to a more detailed model of the specimen.

Using the ABAQUS/Standard package, the x and y components of strain were calculated at each of the loads for which a moiré picture had been collected. One typical pair of strain maps is illustrated below in Figure 6. Knowing the exact location on the specimen of the moiré sequence, the loading conditions were now determined.

A second FE model was constructed in ABAQUS, this time from a micrograph of the specimen's microstructure. The model comprised 6646 elements and about 64000 variables, and was run on Silicon Graphics Octane and Origin computers.

First of all by thresholding, and then using some artistic licence, the areas representing large HMX crystals were defined, and the model was then meshed as illustrated in Figure 6. The large explosive crystals were assumed to be rigid, and the rest of the composition was assumed to be binder. The binder material was assumed to

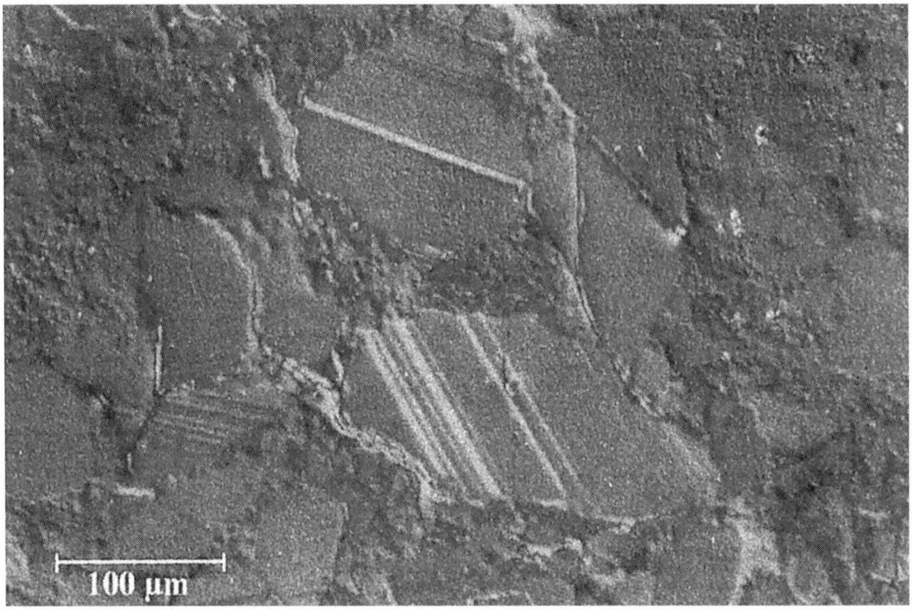

Figure 5. Micrograph of region shown in figure 3.

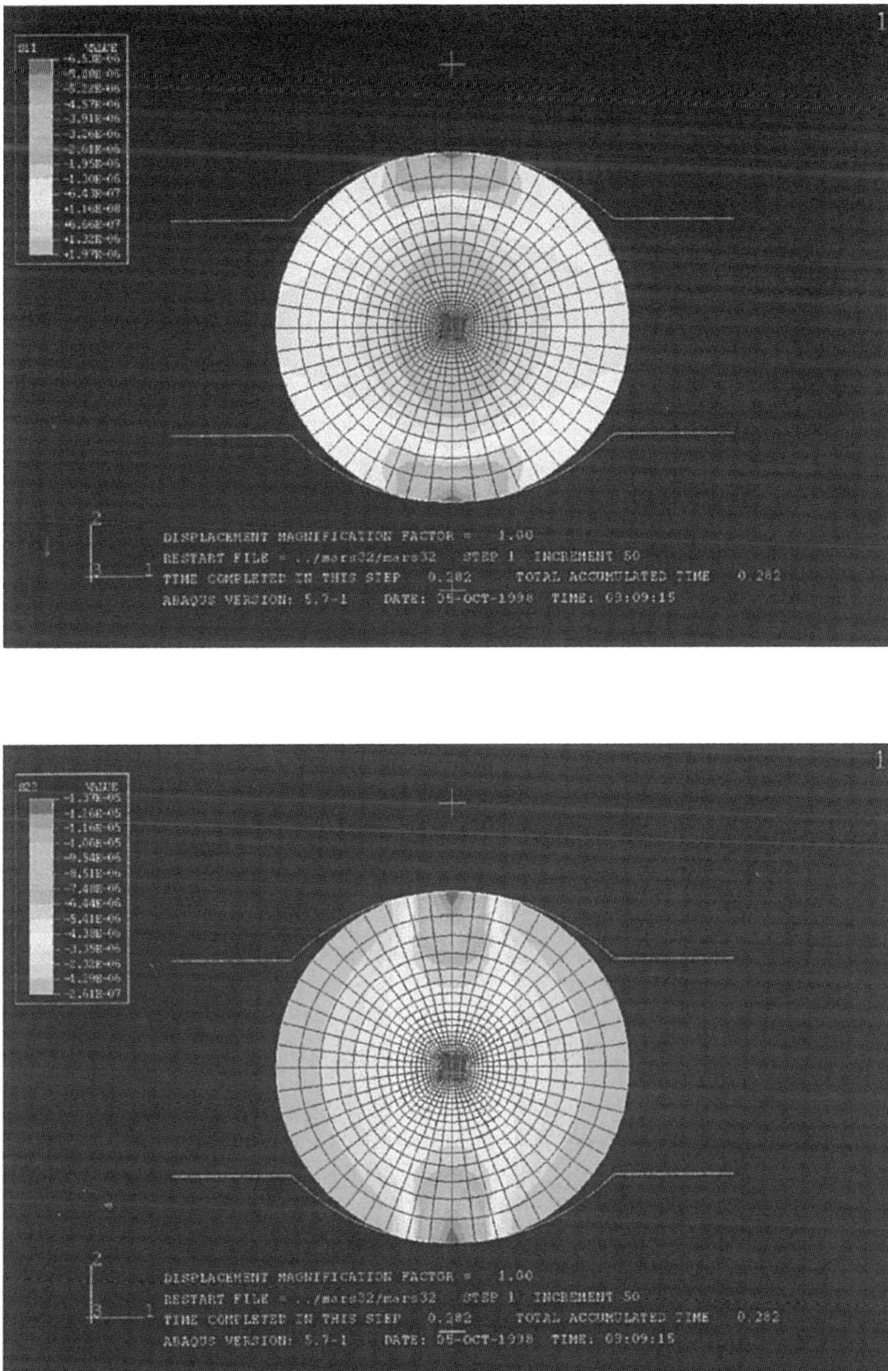

Figure 6. A pair of strain maps of the brazilian test calculated with Abaqus/standard. The upper image represents ε_{XX} and the lower one ε_{YY}.

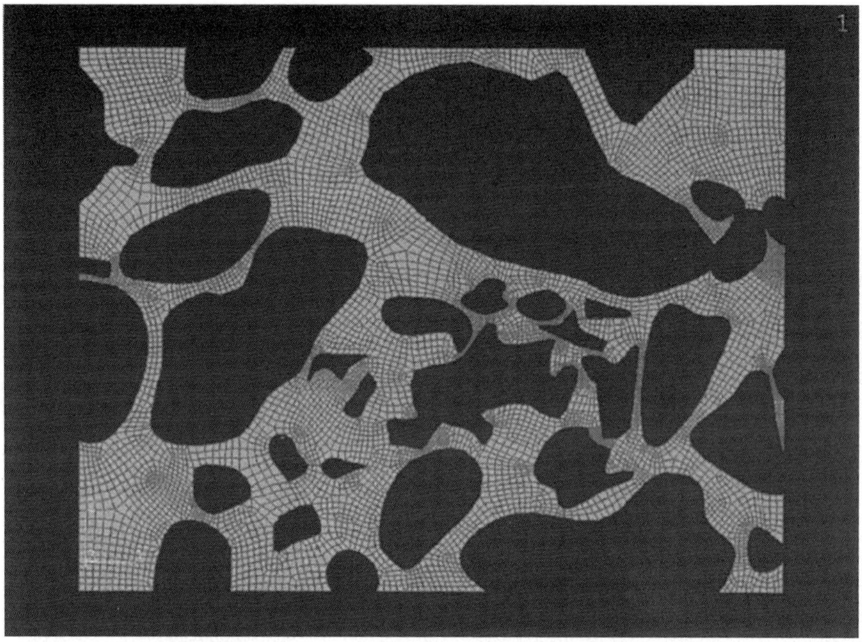

Figure 6. Mesh of microstructural model used in ABAQUS.

behave as an elastomer and any time dependence was ignored, using ABAQUS' "Hyperelastic" model. Applying the boundary conditions previously calculated, strains were calculated, and an example is shown in Figure 7.

Most importantly, as can be seen, the locations of strain concentrations in the model match the measured images well,. This enables some important conclusions to be drawn. First of all, the model has no knowledge of the 3-D structure of the PBX. It is perfectly possible that the large crystals appearing the micrographs are almost vanishingly thin. If this were true, one might expect that the strains measured in the region of such a crystal would not match well the results calculated by this 2-D model which assumes the crystals to be semi-infinite. We have no experimental capability to probe the 3-D structure of the explosive with sufficient resolution to make any appropriate measurements here, and the moiré technique is simply a surface measurement by definition. The quality of the correlation suggests that 3-D effects seem to be insignificant in this case. With this in mind, it was then possible to alter the characteristics of the binder/filler interfaces in the model. The results illustrated in Figure 7 assume a perfectly bonded though freely sliding interface. The influence of assuming a perfectly debonded interface, or even an interface which debonds at a given stress has been investigated. This first experimentally-based FE model of a PBX's microstructure has already proven to be of significant value.

Significant problems remain. Firstly, using the meshing technique described above, roughly 40% of the composition has been assumed to be binder, a figure much higher

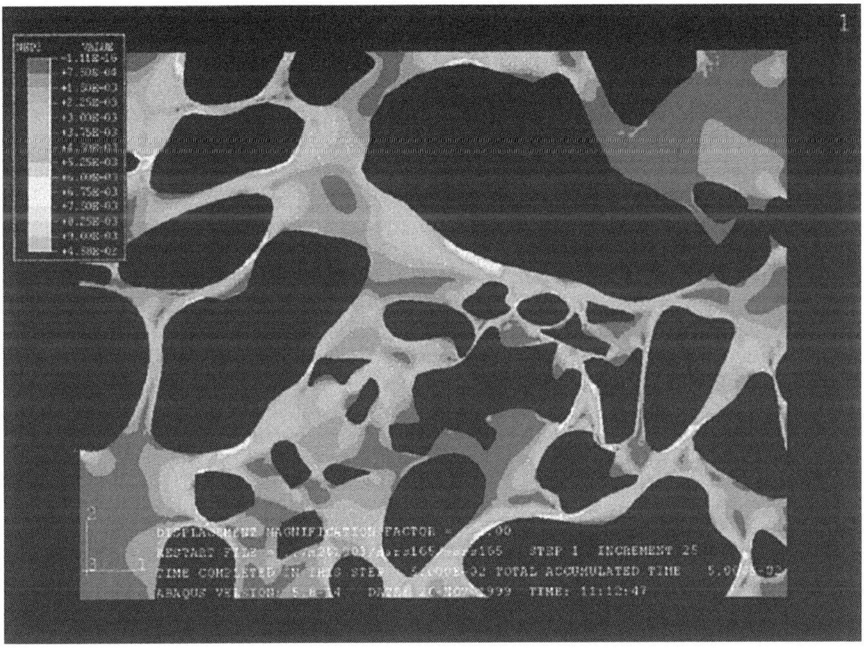

Figure 7. Calculated strains in region corresponding to the moire-measured region.

than is true in fact. In future, it may be appropriate to assign to the binder the properties of a material consisting of a mixture of binder and fine filler particles, rather than pure binder. There is some support for this from separate atomic force microscopy (AFM) studies in the Cavendish[13]. Nonetheless, the accuracy of the model already may suggest that the sensitivity to the presence of fine filler particles is low. The sliding interfaces are not physically based, and this aspect of the model will have to be improved, as well as some means of accounting for strain in the crystalline filler. Whilst the filler has a modulus at least four orders of magnitude greater than the pure binder material in this PBX, experiment shows non-zero strains exist in many explosive crystals, and eventually, the model must reflect this aspect of the experimental results also.

5. CONCLUSIONS

Moiré interferometry may be used to obtain quantitative displacement information about PBXs. These data have been successfully modelled for the first time, allowing model investigations of various parameter studies to be commenced. By combining sensitive optical displacement measuring techniques with a sophisticated model, significant advances in understanding of the failure processes at work in a PBX have been possible. As noted earlier, the developed methods are thought to be of wide applicability

Acknowledgments

The authors wish to thank Prof. J.E. Field for his support and encouragement throughout this research. The research both at Cambridge University and ICI Technology was sponsored by AWE Aldermaston. H.T.Goldrein is also grateful to AWE Aldermaston for support in the form of a William Penney Fellowship. EPSRC and the Royal Society of London are thanked for equipment grants in support of the research.

References

1. Timoshenko SP., Goodier JN., "Theory of Elasticity", Magraw Hill, London, (1970)
2. Awaji H., Sato S., "Diametral Compressive Testing Method", *J. Eng. Mat. Tech.*, Vol. 101, pg. 139, (1979)
3. Johnson HD., "Diametric Disc and Standard Tensile-Test Correlation Study", *Rep. no. MHSMP-81-22*, Amarillo, Texas, Mason and Hanger, (1981)
4. Palmer SJP., Field JE., Huntley JM., "Deformation, Strengths and Strains to Failure of Polymer Bonded Explosives", *Proc. R. Soc. Lond.*, Vol. A 440, pg. 399, (1993)
5. Field, J.E., Parry, M.A., Palmer, S.J.P., Huntley, J.M., "Deformation and Explosive Properties of HMX Powders and Polymer Bonded Explosives", *9th Symposium (International) on Detonation*, Office of Naval Research, Portland, Oregon, Sept 1989, pp. 886–896.
6. Goldrein H.T., Huntley J.M., Palmer S. J. P., Whitworth M. B., Field J. E., "Optical Techniques for Strength Studies of Polymer Bonded Explosives", *10th Symposium (International) on Detonation*, Office of Naval Research, Portland, Oregon, Sept 1993, pp. 525-535
7. P. J. Rae, H. T. Goldrein, S. J. P. Palmer, J. E. Field, A. L. Lewis, "Studies of the failure mechanisms of polymer-bonded explosives by high resolution moire interferometry and environmental scanning electron microscopy", *11th Symposium (International) on Detonation*, Office of Naval Research, Portland, Oregon, Sept 1998, *in press.*
8. Post D., "Moiré Interferometry", Handbook on Experimental Mechanics, ed. A.S. Kobayashi, Prentice-Hall, Eaglewood Cliffs, NJ (1987)
9. Goldrein HT., "Applications of Optical Strain-Measurement Techniques to Composite Materials", PhD Thesis, Pembroke College, Cambridge, (1996)
10. Post D., Han B., Ifju P. "High Sensitivity Moiré: experimental analysis for mechanics and materials" Springer Verlag, NY/London (1997)
11. Palmer S.J.P, Field J.E, "The Deformation and Fracture of β-HMX". Proc. R. Soc. Lond. A383. 1982, pp. 399-407
12. Rodriguez J. et al. "Splitting tests: an alternative to determine the dynamic tensile strenth of ceramic materials." J. Phys IV France Colloq. C8 (DYMAT 94) **4** pg. 101-106 (1994)
13. W.G. Proud, Cavendish Report, Cavendish Laboratory, Madingley Road, Cambridge CB3 0HE (June 2000)

MEASUREMENT OF MICROSCOPIC RESIDUAL STRESS BASED ON THE EVOLUTION OF SURFACE ROUGHNESS DURING SHALLOW CHEMICAL ETCHING

Kyung-Suk Kim
Brown University, Providence, RI 02912, USA

Abstract. A new novel experimental method, surface roughness-evolution spectroscopy (SRES), to measure residual stress components in solids with high spatial resolution is presented. It is based on stress-induced surface roughening caused by shallow chemical etching.

First, a theoretical result of stress-induced roughening in a shallow chemical etching is obtained for isotropic homogeneous solids. The theory predicts that in a shallow chemical etching the roughness grows for a spatial frequency below a stress-dependent critical value, while the roughness of higher frequency decays. The stress dependence of this phenomenon has been found useful for measuring microscopic residual stresses in solids. An AFM (Atomic Force Microscopy) is used to measure the surface-roughness evolution during a shallow chemical etching. The etching depth is typically in the order of a few tens of nano-meters. Experimental results on an aluminum alloy shows that this method is also applicable to anisotropic solids. The typical spatial resolution of the area where the residual stress is measured is 10 by 10 square micro-meters, with a typical stress resolution of 1 MPa in aluminum. Since this method uses the chemical potential, this is a stress gage, not a strain gauge.

This experimental method of skin test for residual-stress measurement is expected to make a significant impact in assessing manufacturability, reliability and residual life of solid components, which is essential in developing and maintaining proper infrastructures of civil and mechanical systems.

Reference

K.-S Kim, K.-S., Hurtado, J. A., Tan, H., Physical Review Letters, Vol. 83, No. 19, p. 3872-3876 (1999)

P. Ståhle and K.G. Sundin (eds.), IUTAM Symposium on Field Analyses for Determination of Material Parameters - Experimental and Numerical Aspects, 11.

THE INCUBATION TIME - THEORY AND EXPERIMENT

N. F. Morozov and Y. V. Petrov
Faculty of Applied Mathematics, St. Petersburg State University
Bibliotechnaya 2, St.Petersburg, 198904 Russia

Abstract. We are proposing a clear definition of the term "incubation time" and we show how it can be experimentally and theoretically determined. We also demonstrate the nonhomogeneity parameter of a material and the connection between incubation time and thermodynamical characteristics.

1. INTRODUCTION

For the past 50 years engineering practice and scientific work have dealt with high-rate loading in micro- and nanosecond range. Attempts to extend of static and quasi-static fracture criteria for the prediction of fracture were unsuccessful. We note the mental experiment by Cherepanov[1], well known experiments by K. Ravi-Chandar, W. Knauss[2], J. Dalley[3], D. Shockey) and L. Seaman[4,5]. It is necessary to propose, to discuss and to test the conditions of dynamic fracture. For that the material parameters, which define its dynamic strength are needed.

Such a parameter, although not clearly defined, was proposed by J. Kalthoff, D. Shockey *et al* [6]. The authors associated this parameter with the minimal interval needed to prepare for fracture and called it *induction time*.

In papers by N. Morozov, Yu. Petrov and A. Utkin[7-9] a criterion of dynamic brittle fracture with a new parameter — incubation time τ defined by

$$\max \frac{1}{\tau} \int_{t-\tau}^{t} dt' \int_{0}^{d} \sigma_{\mathrm{I}} dr = \sigma_c . \tag{1}$$

where

$$d = \frac{2}{\pi} \frac{K_{\mathrm{Ic}}^2}{\sigma_c^2} . \tag{2}$$

To determin e τ we propose[10] the limiting amplitudes method. According this method we should fix the loading duration T, and increase the loading amplitude Π

P. Ståhle and K.G. Sundin (eds.), IUTAM Symposium on Field Analyses for Determination of Material Parameters - Experimental and Numerical Aspects, 13–16.

until fracture occurs. Then we change the loading duration and repeat the experiment. So, for a series of specimens we obtain the function

$$\Pi = \Pi(T). \tag{3}$$

As T goes to zero, Π goes to infinity such that

$$U = \Pi T \tag{4}$$

attains a constant value. The criterion (1) determines the delay of fracture t_*; it is related to the incubation time τ by

$$t_* = \frac{U_0}{U_*}\tau. \tag{5}$$

Here U_0 is the threshold instantaneous force, U_* is the instantaneous force applied to the specimen.

Experimental data show that fracture delay t_* takes constant values for constant value of τ for small but not zero values of T. So, we can determine τ from laboratory measurements.

The incubation time τ is considered to be the only known parameter that defines the dynamic strength at high-rate loading. We propose to introduce a practical test to determine τ.

The determination of the fracture parameters d and τ is an essential condition for successful using of the criterion (1). The method of parameters determination was described above. The parameter d for laboratory specimens is defined uniquely by (2), but τ is not. The experiments and calculations show that τ is depending on whether the crack tip is in a defect-free region of the specimen ($\tau = \tau_i$) or near a stress concentrator ($\tau = \tau_c$) (see Table 1).

Table 1. Incubation time for different regions of material

Problems of small fracture and erosion		Problems of crack start	
Metal	Polymer	Metal	Polymer
$\tau_i \approx 0.5 - 1.0$ μs	$\tau_i \approx 0.5 - 1.0$ μs	$\tau_c \approx 10 - 14$ μs	$\tau_c \approx 18 - 20$ μs

It becomes necessary to identify yet another critical value, τ_g, of τ. This applies when the fracture occurs in the neighborhood of a smooth boundary. To determine the relationship between values of structural time, τ, for different regions in a material is an challenging problem in the study of strength of materials.

Let us make the following assumption: the relationship between τ_i, τ_g and τ_c is similar to relationship between period of vibrations of Debye oscillators T_i^D, T_g^D and T_c^D [11]. Here indexes i, g and c signify the intact, boundary and defect region of the same material.

Fix the value of T_i^D and evaluate the value of T_g^D and T_c^D with respect to it.

Let us take the frequencies of Debye oscillations

$$\nu_c^D = \frac{2\pi}{T_c^D}, \quad \nu_i^D = \frac{2\pi}{T_i^D}, \quad \nu_g^D = \frac{2\pi}{T_g^D}. \tag{6}$$

According to Gruneizen's law[11] we have

$$\frac{\nu_i^D}{\nu_c^D} = \left(\frac{V_c}{V_i}\right)^{\gamma}, \tag{7}$$

where γ is a material parameter, $\gamma \in (0, 1)$, and V_c and V_i are character volumes of these regions.

Near the crack tip due to the deformations singularity we have

$$\frac{V_c}{V_i} \to \infty. \tag{8}$$

Therefore,

$$\nu_c^D \to 0, \quad T_c^D \to \infty, \quad T_c^D > T_i^D. \tag{9}$$

By virtue of our assumption we have

$$\tau_c > \tau_i. \tag{10}$$

It is known that amplitude of Debye oscillations near a free boundary is greater than in volume, so for estimation of its frequency ν_g^D the nonlinear terms should be taken into account. In continuum mechanics, the anharmonic vibrations are described frequently by Duffing equation

$$\ddot{x} + \omega^2 x + \mu x^3 = 0. \tag{11}$$

By using the Ritz- Galerkin method or Krylov – Bogolyubov harmonic balance method we obtain

$$\omega^2 = \omega_0^2 + \mu c^2 \tag{12}$$

and

$$T = \frac{2\pi}{\omega} = \frac{2\pi}{\omega_0 \sqrt{1 + \mu c^2 / \omega_0^2}} \approx \frac{2\pi}{\omega_0}\left(1 + \frac{1}{2}\frac{\mu}{\omega_0^2}c^2\right) = T_0\left(1 + \frac{1}{2}\frac{\mu}{\omega_0^2}c^2\right), \tag{13}$$

that is

$$\tau_g \approx \tau_i(1 \pm \alpha), \tag{14}$$

where α have to be determined for each specific load an specimen geometry.

In the view of our assumptions, formulae (10) and (14) now give the relationships between incubation time values for different regions of investigated materials.

References

1. Cherepanov, G.P. (1974) Brittle fracture mechanics. Nauka, Moskow (in Russian).
2. Ravi-Chandar and K. Knauss, W.G. (1984) An experimental investigation into dynamic fracture, Int. J. Fracture, 25, pp. 247-262.
3. Dalley, J.W. (1979) Dynamic photoelastic studies of fracture, Exp. Mech., 19, pp. 349-361.
4. Shokey, D. et al. (1983) Application of microstatical fracture..., Material behavior under high stress and ultrahigh loading rates, New York, pp. 273-293.
5. Seaman, L. et al (1985) A continuum model ..., J. Appl. Mech., 52, pp. 593-600.
6. Shokey, D. et al (1986) Short-pulse fracture mechanics, Eng. Fracture Mech., 23, pp. 311-319.
7. Morozov, N.F. and Petrov Yu.V. and Utkin A.A (1988) On limit intensity computation..., News Russian Science Acad. Solids Mech., No. 5, pp. 180-182 (in Russian).
8. Morozov, N.F. and Petrov Yu.V. and Utkin A.A (1990) On cleavage analysis..., Report Science Acad. USSR, 313 (2), pp. 276-279 (in Russian).
9. Morozov, N.F. and Petrov Yu.V. and Utkin A.A (1988) On fracture near crack tip..., Physical-Chemical Material Mechanics, No. 4, pp. 75-77 (in Russian).
10. Morozov, N.F. and Petrov Yu.V. (1994) On the modeling of fracture, J. Appl. Mech., 61, pp. 710-712.
11. Gerihalco L. (1973) Statistical physics of materials, Wiley.

AFFORDABLE TMF TESTING SUPPORTED BY FE SIMULATIONS

Magnus Hasselqvist
Alstom Power Sweden AB, SE612 82 Finspong, Sweden

Abstract. The predictive capability of cyclic life assessment methodology for gas hot parts of turbines is analysed and improved.

The main finding is that standard test data is seldom sufficient to guide the modelling,or to calibrate the resulting model when one aims for predictive capability under component conditions. We therefore focus on improved testing as a basis for improved modelling.

Constitutive characterisation tests have been performed and modern state variable models are evaluated. We find that there are models that can be used with some modifications, for particle strengthened blades and vane materials such as IM792, but no adequate models for solution-strengthened combustor materials such as Haynes 230. Constitutive models for IN792 are being developed and calibrated using creep data and step relaxation tests tailor made for robust parameter estimation of back stress evolution models.

Blades such as IN792 have high temperature strength, but must be coated to survive the corrosive environment. For base metal/coating aggregates, it has been recognised that TMF tests are needed, i.e. tests in which strain and temperature aresimultaneously varied. The current TMF test methodology is toprovide a standard LCF test rig with a variable temperature oven. However, the resulting test rig becomes difficult to handle and the difference between test and component conditions is in many cases still too large. Work is on-going to develop an easy-to-handle TMF test rig which allow one to obtain realistic conditions. The method require extensive FE simulations to compute material parameters and evaluate results.

1 MODELLING OF THE COMBUSTOR MATERIAL HAYNES 230

Haynes 230 is a nickel base superalloy which derives resistance to inelastic deformation from solution and carbide strengthening. It contains primary carbides when delivered from the manufacturer. Secondary carbides may or may not precipitate in use depending on temperature and straining. It is ductile and hot corrosion resistant. More information and further references can be found in Hasselqvist [5].

In the test CR0113, fast cycling, triggering carbide precipitation through creation of initiation sites in the form of dislocation networks, was periodically interrupted as seen in Figure 1. Note that the stress at the end of the relaxation decrease with increasing strain rate in the preceding ramp, and that the strain rate dependency is

P. Ståhle and K.G. Sundin (eds.), IUTAM Symposium on Field Analyses for Determination of Material Parameters - Experimental and Numerical Aspects, 17–25.

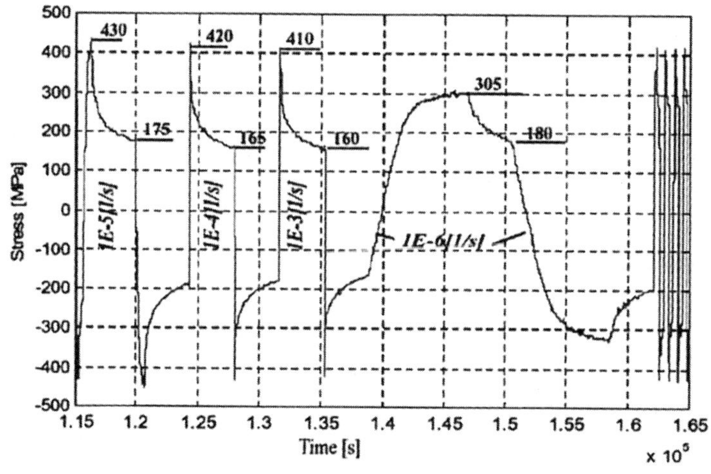

Figure 1 Close-up of a second block in characterisation test CR0113 at 700 °C

negative above 1E-5[1/s]. The stress levels change with the block number. The stress at the end of relaxation is for example consistently lower in the fifth block for which the carbide content is higher, see Figure 2.

1.1 Constitutive testing/modelling

Our interpretation of the results in Figure 1 is that the solute drag, i.e. strengthening through pinning of dislocations by Cottrell clouds of solutes, is reduced in high strain rate ramps. This effect is evidently strong enough to counter the positive strain rate dependency of grain boundary sliding and dislocation climb (the likely deformation mechanisms) to give an overall negative strain rate dependency.

Carbide precipitation provides particle strengthening, but the amount of solutes and hence the solute drag decreases. The strength at high strain rates, where solute

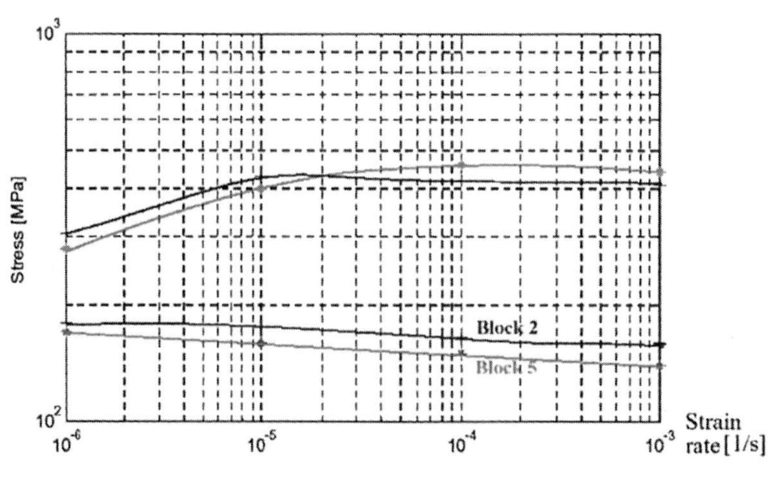

Figure 2 Upper lines: Maximum stress (upper lines) and stress at the end of the hold time (lower lines) for the second and fifth block in CR0113.

drag is never effective, increases but the strength at low strain rates decreases since particle strengthening is not enough to counter the loss of solute drag, see Figure 2. Tests at lower temperatures show an even stronger negative strain rate dependency starting at even , since their state variables are typically back stress and work hardening, whereas the deformation behaviour in Haynes 230 depends more on dynamic variations in solute drag and carbide morphology.

In the end, we were forced to resort to standard Norton modelling with various correction factors chosen so that inelastic strain ranges are predicted fairly well for loading rates typical for Alstom Power Sweden gas turbines (i.e. we have a narrow but sufficient domain where the model can be applied).

The interpretation of test data is based on Cahn&Haasen [4], [7], [13] and Hull&Bacon [8]. Overviews of modern state variable modelling can be found in Chaboche[1], Miller [9], and Stouffer & Dame [12]. The author has found one model incorporating solute drag, see Miller [10].

1.2 Life testing/modelling

The work was focused on predictive capability under In-Phase Thermo Mechanical Fatigue (IP TMF), see Fig. 3. This damage case was considered the most challenging since thecombination of cyclic straining and tensile hold at high temperature implied that interaction between cyclic fatigue and creep cavitation was possible.

The standard procedure for cyclic life estimation is to use the linear damage summation rule (codified in ASME N-47) calibrated with creep-rupture and LCF data to predict the number of cycles N to crack initiation. N_f is the number of cycles to crack initiation in a standard LCF test done at strain range Δ϶. t_r is the time to rupture in a creep test done at the stress σ.

Fatigue damage per cycle $$D_f = \frac{1}{N_f(\Delta\varepsilon)}$$

Creep damage per cycle $$D_c = \int_{\text{hold time}} \frac{dt}{t_r(|\sigma|)} \qquad (1)$$

Crack inititation criterion $$D = N(D_f + D_c) = 1$$

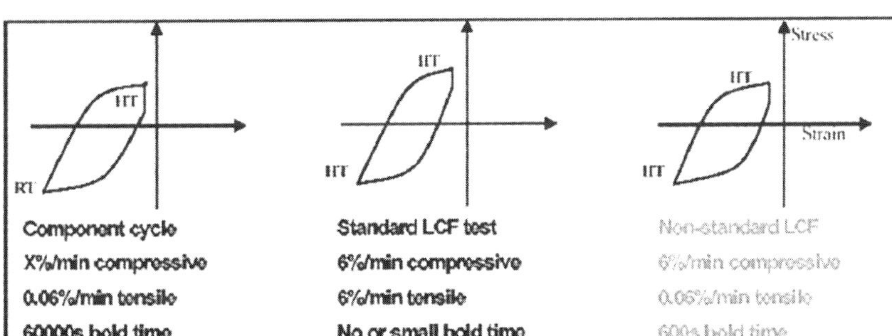

Figure 3 Left - Typical IP TMF component cycle. Middle - Standard LCF test. Right - Non-standard LCF test designed to get 'as close to the component cycle as it is possible to get in a standard LCF test machine'.

The model assumes 'plasticity', i.e. strain rate independence, under ramps, and 'creep', i.e. strain rate dependence, under hold times. It is furthermore assumed that the 'plasticity' leads to 'fatigue' whereas 'creep' leads to 'creep damage'.

Numerous variants exist, some of which require LCF tests with hold time:

- The fatigue term may be formulated in terms of the 'plastic' or inelastic strain range rather than the total strain range
- A weight function may be added to the integral to suppress accumulation of creep damage under compressive loading
- The integral may be replaced with a simpler expression describing the ductility exhaustion

Material data show that the assumption of 'plasticity' during ramps and 'creep' during hold times' is not justified for Haynes 230 at 800°C (the author has in fact not seen any material for which the assumption makes sense at any temperature). Therefore all models based on this assumption were excluded. Furthermore, the use of total strain range is questionable, since the hold time increases the inelastic strain range and hence should lead to more fatigue damage. We therefore decided to base the fatigue damage model on inelastic strain range.

The creep-rupture integral in the linear damage summation rule is not meaningful for the ductile material Haynes 230, since the stress is far from constant during the test (assuming standard load controlled testing).

The main alternative is a ductility exhaustion model, i.e. the same accumulated creep strain is allowed in the component as in creep tests. But analysis of creep data, taking the reduction in cross section area into account, reveals that what is perceived as "tertiary creep" is actually secondary creep under increasing stress, suggesting that creep specimens break due to geometrical instability rather than due to cavitation.

This suggests that creep damage, in the sense of cavitation, does not occur until the material has accumulated much more inelastic strain than in creep tests. Use of creep data would therefor lead to over-conservatism, i.e. expensive and overly complicated design solutions. We conclude that we need something else than creep data to evaluate creep damage. Let us therefor try to simulate reality.

The real component cycle, seen on the left in Figure 4, is poorly represented by the LCF test in the middle. The LCF test was however modified so that the tensile going strain rate was reduced to the same rate as in a gas turbine but the compression going ramp rate was kept high. A hold of 10 min. is long enough that the difference between the test and the component cycle in terms of inelastic strain accumulation during the hold time is very small. Creep cavities should then grow (in tension) in the same manner in the test as in the component. Since the component is in compression only at low temperature, cavities will not shrink. A fast compressive rate in the test will also suppress shrinkage, thereby simulating the effect of room temperature loading. We performed a few tests of this type.

Figure 4 shows a plot of cyclic life v. inelastic strain range (for the stabilised cycle) for standard and non- standard LCF tests. The correlation is remarkable. It appears that no creep damage term is needed. Whether this holds also for other test temperatures, for multiaxial loading and under anisotropic cycling remains to be seen. We are continuing to test.

The analysis above was based on Davis [2], Suresh [14] and Riedel [11].

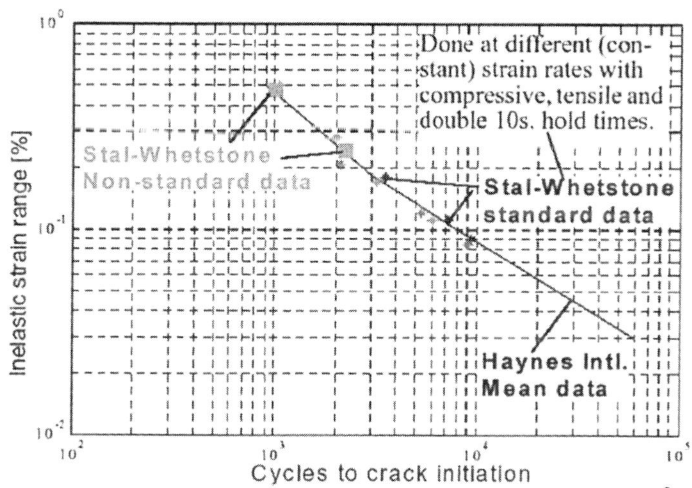

Figure 4 Standard and non-standard LCF data for Haynes 230 at 800 ° C

2 MODELLING OF THE BLADE MATERIAL IN792

The particle strengthened material IN792 has high strength at elevated temperature but is not very ductile. It is almost always used with a coating to protect against hot corrosion and oxidation. We need the ability to predict the redistribution of stress and strain in the first cycle (which should result in stabilised loops that are virtually elastic except at severe notches) and to evaluate if the base metal / coating combination has a long enough cyclic life.

2.1 Constitutive testing/modelling

This work has its roots in experience gained during a single crystal constitutive modelling project, see Hasselqvist[6]. In that project we tried to calibrate a model of Chaboche type (many of parameters, back stress and work hardening as state variables) using standard material data. The outcomes of that project were as follows:

- Standard test data often exhibit phenomena (solute drag associated softening/ hardening, incubation time prior to primary creep, particle coarsening leading to tertiary creep etc.) not associated with back stress and work hardening.
- The combination of
  - many material parameters
  - the discrepancy between mechanisms acting in tests and in the model
  - the low information content in standard tests led to arbitrariness in parameter values despite an earnest attempt to perform the calibration in a step-by-step manner and to interpret the parameters at each step.
- The few constitutive tests we had at the time (we had just started to explore the possibility of non-standard testing) made the calibration much more robust in the sense that the minimisation process did not find (as many) sub-minima.

We concluded that successful modelling will require

- much more reliance on non-standard testing

- a ranking of deformation mechanisms w.r.t. significance under component conditions to guide the modelling/testing. This will require
- better knowledge about component conditions
- elimination of standard tests that trigger mechanisms that are not significant under component conditions to avoid too-complex models

Characterisation testing on IN792 revealed that the most significant mechanisms under component conditions are back stress, particle coarsening, see Figure 5, solute drag and work hardening (in that order).

The testing/modelling work, which is done together with Chalmers University, starts with models using "only" back stress as state variables and is now proceeding with a (crude but useful) modification to take particle coarsening into account. Essentially, we let the low temperature yield strength be a function of the maximum cyclic temperature since this, and the exposure time, govern the microstructure, see Figure 5.

The creep data available for IN792 had not been sampled very well and did not allow us to extract more than the secondary creep rate as a function of stress. High temperature LCF and tensile tests had been done but, as usual, at strain rates far outside the domain of interest w.r.t. conditions in our blades. Some further characterisation tests indicate that these high strain rate tests would "contaminate" the calibration process as they would trigger a 'solute drag mode' not seen in components or creep tests.

Instead, we started to perform step relaxation tests which give good information about back stress evolution under component conditions, see Figure 6. These tests are used together with secondary creep data for calibration of models. The tests are done at Alstom Power Sweden. The numerical work, parameter optimization and time integration of the ODE system describing the model, is done at Chalmers by Ekh who used this work in his dissertation [3].

Even though the project is still in progress, and additional tests as well as modifications to the models and the numerical implementation are necessary, we can conclude that the use of constitutive tests for characterisation and robustness in calibration is highly useful.

2.2 Life testing/modelling

For base metal / coating aggregates, it has been recognised that TMF tests, i.e. tests in which strain and temperature are simultaneously varied, are needed, see [6].

The current TMF test method is to fit a standard LCF rig with a variable temperature oven, the resulting test rig is very complex to handle and the difference between test and component conditions is in many cases still too large. The time for a TMF test is for example very short compared with component conditions. The layered structure will change in service and might eventually result in a structure with a very poor TMF resistance (see Fig. 7). Therefore a short term TMF test might lead to a non-conservative life estimate. Although the stress state (in the base material) is uniaxial while many important TMF loadings are multiaxial (see Fig. 8).

Work is progressing to develop simple TMF tests rigs which allow us to get closer to component conditions. One of these is the 'hot spot' rig, which relies only on thermals training obtained through heating, cooling and insulation (i.e. the same way in which we get TMF in a gas turbine).

In this rig we reproduce a damage case in which a small spot is significantly hotter than its surroundings. Typical figures are 950 o C in the spot and 800° C around it. This produces additional thermal mismatch, on top of the mismatch due to

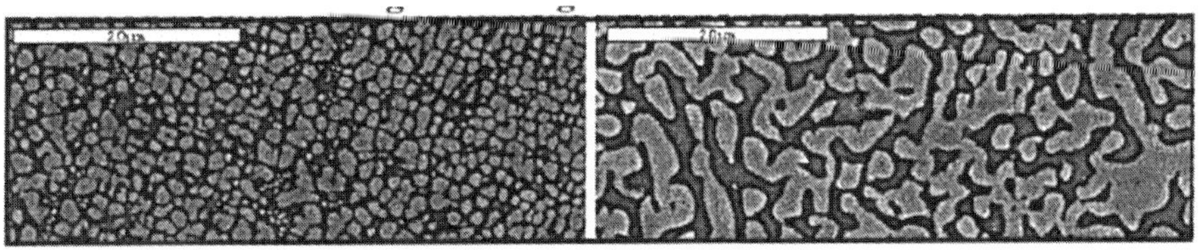

Figure 5 Left: The γ' particle mean diameter is 1.55μm. Right - Rafted micro-structure. From serviced blades. Initial mean diameter was 0.8μm.

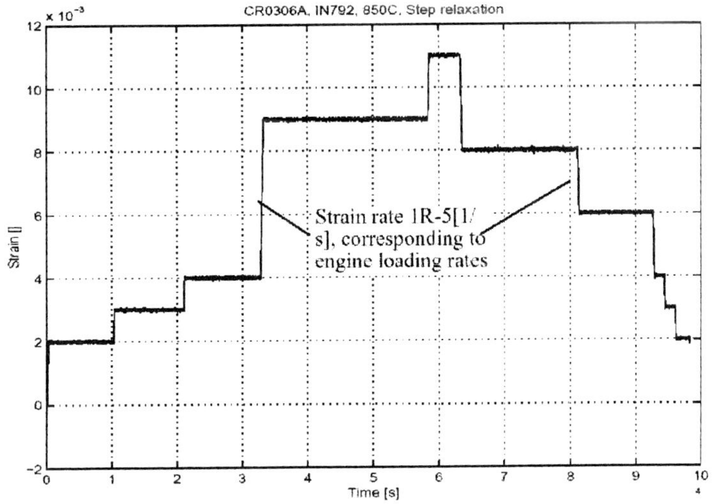

Figure 6 Strain history for part A of CR0306

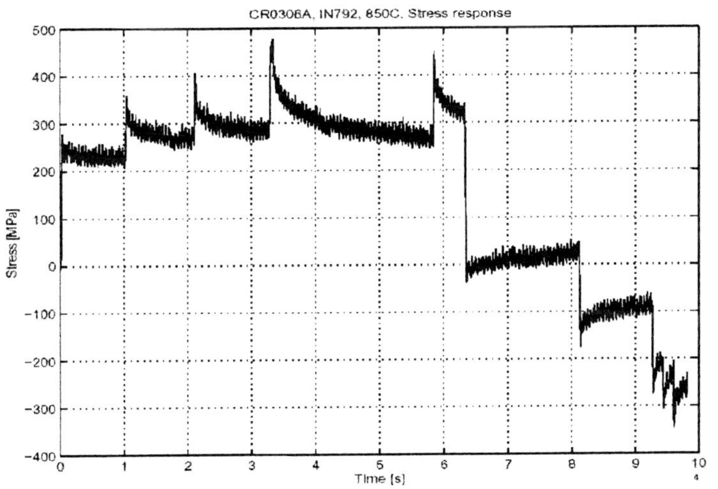

Figure 7 Stress response for part A of CR0306

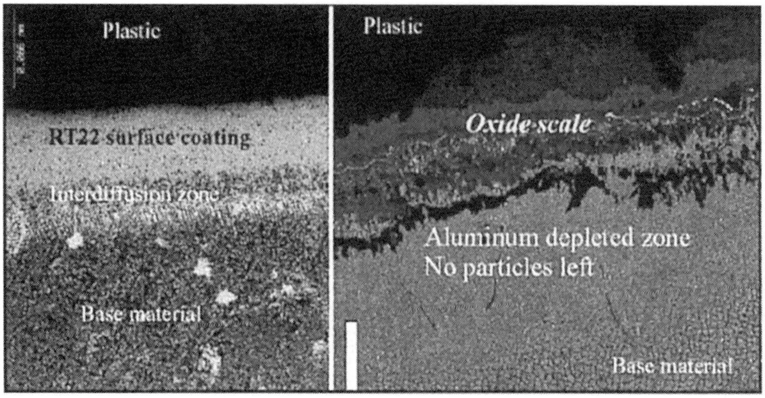

Figure 8 Left - Surface protected by RT22 coating. Right - "mushy" oxide scale and • ' depletion of an unprotected surface. Note that there is always a coating, whether it is applied or the materials "own" oxide scale.

differences in thermal expansion between the base metal and the coating. Depending on the signs of these mismatches, the hot spot can increase or decrease the strain range in the coating. It makes sense to test a number of base metal / coating combinations, and choose one which can withstand a reasonable temperature difference in order to have robustness against hot spots.

Figure 9 shows a schema for the prototype test rig. It will be easy to use and cheap to run but, while stress and strain (in the base metal) are measured in current TMF rigs neither of these are measured in the new test rig. They have to be calculated. Planning and evaluation of tests with the new rig will require a transient temperature calculation as input to a stress calculation since the hot spot is almost in strain control, the constitutive model needs to be merely adequate, not precise.

A heat transfer model, based on understanding of radiation and fluid mechanics, will be calibrated using a thermo camera and thermo-couples. CFD calculations will be employed during the final design stages and plastic mock-up's for flow tests might be necessary to back those up.

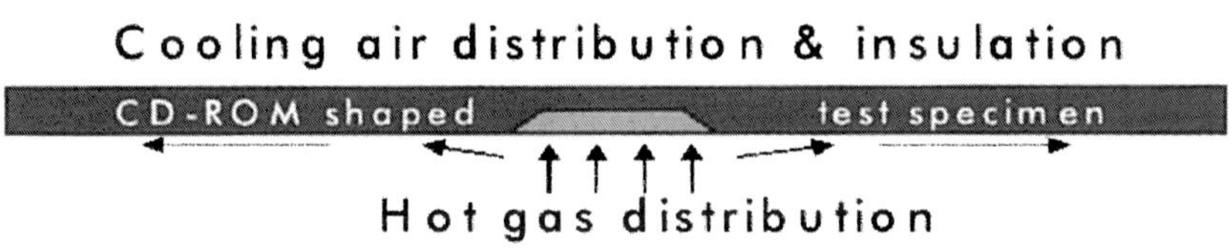

Figure 9 Suggested test rig for 'hot spot' TMF testing

References

[1] J. L. Chaboche Unified Cyclic Viscoplastic Constitutive Equations: Development, Capabilities and Thermodynamic Framework - Part of 'Unified Constitutive Laws of Plastic Deformation' Academic Press 1996

[2] J. R Davis(Ed.) ASM Specialty Handbook - Heat Resistant Materials ASM 1997

[3] M. Ekh Modelling of Inelastic Response of Metals with Emphasis on Cyclic Viscoplasticity - Dissertation 1572 (new series) at Chalmers University of Technology Chalmers Reproservice 2000

[4] P. Haasen Mechanical Properties of Solid Solutions - Ch.23 in Physical Metallurgy, 4th ed. vol.3 North-Holland 1996

[5] M. Hasselqvist Differences in Morphology Between LCF-Cycled, Crept and Thermally Exposed Haynes 230 and a Discussion about the Implications for Constitutive and Life Modelling for Real Component Conditions – Part of 'Modelling of Microstructural Evolution in Creep Resistant Materials' University press, Cambridge, 1999

[6] M. Hasselqvist Life Assessment of Single Crystal Internally Cooled Blades with Emphasis on Thermo-Mechanical Fatigue - A Report for GTC RT RGS 8/99, ABB Alstom Power Sweden

[7] J. P. Hirch Dislocations - Ch.20 in Physical Metallurgy, 4th ed. vol.3 North-Holland 1996

[8] D. Hull, D. J. Bacon Introduction to Dislocations 3rd Ed. Pergamon 1984

[9] A. K. Miller The MATMOD equations - Part of 'Unified Constitutive Equcations for Creep and Plasticity' Elsevier 1987

[10] A. K. Miller et al. Improvements in the MATMOD equations for Modelling Solute Effects and Yield-Surface Distortion - Part of 'Unified Constitutive Laws of Plastic Deformation' Academic Press 1996

[11] H. Riedel Fracture at High Temperatures Springer Verlag 1987

[12] D. C. Stouffer, L. T. Dame Inelastic Deformation of Metals - Models, Mechanical Properties and Metallurgy John Wiley & Sons 1996

[13] J. L. Strudel Multiphase Alloys - Ch.25 in Physical Metallurgy, 4th ed. vol.3 North-Holland 1996

[14] S. Suresh Fatigue of Materials, Cambride University press 1991

IDENTIFICATION OF DYNAMIC PROPERTIES OF A RAILWAY TRACK

V.L. Markine, A.P. de Man, C. Esveld
Section of Road and Railway Engineering Faculty of Civil Engineering, Delft University of Technology Stevinweg 1, NL-2628, CN Delft, The Netherlands

Abstract: The paper presents a methodology for assessment of dynamic characteristics of a railway track in application to an Embedded Rail Structure. First, the dynamic responses of a track are obtained using a Hammer Excitation Test (HET) by applying a vertical impulse load to a rail and measuring resulting accelerations of a track. The measured data is then analyzed using the Frequency Response Function (FRF). The hammer test is also simulated by the finite element model (RAIL), which can be used for analysis of the dynamic behavior of a track under various loading. Both HET and RAIL are developed at Delft University of Technology (DUT). To match the simulation and measurement data a numerical optimization technique called a Multipoint Approximation based on Response Surface fitting (MARS) method has been used. The parameters of the model related to the material properties of elastic compound (polyurethane and cork mixture) wherein the rails are embedded have been varied during the identification process. The numerical model with the obtained parameters of compound can then be used for estimation of performance characteristics of the railway track. The compound parameters have been compared with the ones obtained using a single degree of freedom mass-spring system. The results of the identification problems are presented and discussed.

1. INTRODUCTION

One of the important characteristics of a railway track is the vertical stiffness. Under a high frequency loading the vertical stiffness of a track defines the level of track

P. Ståhle and K.G. Sundin (eds.), IUTAM Symposium on Field Analyses for Determination of Material Parameters - Experimental and Numerical Aspects, 27–36.

vibrations and wheel-rail contact forces, which ultimately define acoustic noise emission and effort on maintenance of rail and wheels. Additionally, large variations in track elasticity can cause serious damage to a track that results in problems on operational safety of a track. Insufficient vertical stiffness of track can yield rail fatigue problems, whereas too great track stiffness leads to high emission of noise, and wear of wheels and rails. That is why assessment of vertical stiffness of a track is extremely important for track maintenance.

Recently, a method for estimation of the dynamic characteristics of a track has been developed within the Dynatrack project [5] at Delft University of Technology (DUT). The procedure includes measurement of dynamic responses of a track, numerical modeling of track behavior, and parameter identification of track components. The dynamic characteristics of a track are obtained using a Hammer Excitation Test (HET) according to which a vertical impulse load is applied to the top of a rail. The resulting track vibrations are used for prediction of noise emission and wheel-rail interaction forces by analyzing the Frequency Response Functions of the measured data.

In the present paper the dynamic properties of a special type of railway track, a so-called Embedded Rail Structure (ERS) are investigated. ERS belongs to a ballastless type of track that has a number of advantages over to a classical track; the most important is low maintenance cost, especially in high-speed operations. A typical ERS applied so far consists of a continuous reinforced concrete slab resting on a concrete stabilized roadbed, which in turn is placed on a sand base (*Figure 1*a). Rails in such a structure are continuously supported by an elastic compound (mixture of polyurethane and cork) as shown in *Figure 1*b. To analyze the dynamic behavior of a track a Finite Element model implemented in a computer program RAIL has been developed at DUT [2]. The model can analyze various track structures under different loadings such as a moving train or temperature variation. Accuracy of track responses obtained from the numerical model strongly depends on values of the model parameters related to track components. Some of these parameters such as rail geometry, are known to a high accuracy; other parameters can only be obtained using a numerical identification technique.

In the identification procedure material properties of the compound such as stiffness, damping and mass density are determined by matching the excitation hammer measurement data with results of numerical simulation. The identification of the material parameters is formulated as a minimization problem. To solve this, we used an optimization technique called a Multipoint Approximation based on Response

Surface fitting (MARS) method [5,8]. The method has been specifically developed for the problems where multiple response analyses and (time consuming)

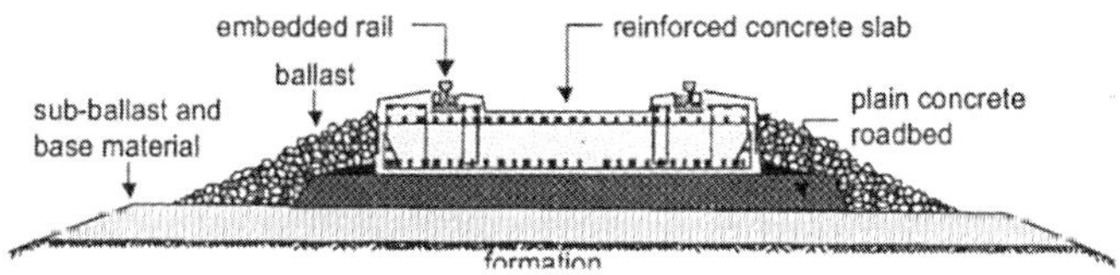

a

b

Figure 1. ERS with conventional (UIC54) rails

simulations are involved. For comparison, the material properties of the compound are also obtained using a single degree of freedom mass-spring system with damping.

2. HAMMER EXCITATION TEST

Measuring track performance in a continuous way (such as e.g. using a High Speed Deflectograph [7]) is essential to obtain information on overall quality of a track, as well as to determine location of 'weak spots'. Usually, such an overall view provides sufficient data to perform maintenance on certain sections of a track. However, sometimes weak spots and, in extreme cases, an entire track require closer examination, e.g. if track problems occur repeatedly. For that purpose a spot measuring method such as a Hammer Excitation Test has been developed [4,5]. Similar to the High Speed Deflectograph, the Hammer Excitation Test concentrates on track elasticity in the vertical direction, but causes a high frequency loading, which more closely resembles noise, vibrations and wheel-rail interaction forces
The major part of track performance that can be assessed by the Hammer Excitation Test, is the dynamic (vibration) behavior of a track (*Figure 2*). The overall performance of a track is strongly defined by characteristics of its components. The analysis of the dynamic behavior of a track includes measurements of an applied load as well as resulting vibrations of track. A vertical (impulse) load applied to the top of the rail initiates track vibrations in the frequency range between 30 Hz and 3 kHz. The vibrations are recorded using accelerometers, which have an application range between 10 Hz and 10 kHz. The recorded signals are then sampled at 2 kHz and converted via Fast Fourier Transform (FFT) into Frequency Response Functions (FRF), we assume a linear behavior for a track. Here, the inertance frequency response function H_{AF} $[kg^{-1}]$ has been used, which relates the applied force (input) F and recorded accelerations of a track (output) A :

$$H_{AF}(f) = \frac{S_{AF}(f)}{S_{FF}(f)} \tag{1}$$

where

Figure 2. Hammer Excitation Test on classical track and Embedded Rail Structure

S_{AF} $[mN/s]$ is the complex cross-spectrum of accelerations (output);

S_{FF} $[N^2 s]$ is the power spectrum of the impulse force (input);

f $[Hz]$ is the frequency.

On the other hand, one can analyze a receptance frequency response function H_{XF} $[m/N]$. Similar to an inertance FRF, H_{XF} relates the applied force (input) F and resulting displacement (output) X

$$H_{XF}(f) = \frac{S_{XF}(f)}{S_{FF}(f)} = \frac{1}{(2\pi f)^2} \cdot \frac{S_{AF}(f)}{S_{FF}(f)}, \qquad S_{XF}(f) = (2\pi f)^{-2} S_{AF}(f) \tag{2}$$

where S_{XF} $[mNs]$ is the complex cross-spectrum of the resulting displacements. From [1]) and (2) the following relation between H_{AF} and H_{XF} can be obtained:

$$H_{AF}(f) = (2\pi f)^2 H_{XF}(f). \tag{3}$$

numerical model

In order to better understand measurement data, we need a numerical model able to describe various behavior of a railway track. Such a model that uses the Finite Element formalism has been developed at the Railway Engineering Group (DUT) and implemented in the computer software RAIL. The choice of the parameters of the model is crucial in assessment and prediction of track performance. Once determined on the basis of measurement data, the parameters can be used in the model for dynamic analysis of various track loadings such as moving train and temperature variation.

The numerical model implemented in RAIL can be used for modeling ballasted as well as ballastless track structures. Depending on a track structure under consideration a numerical model can contain a number of layers. Each layer is a beam on an elastic foundation. A numerical model of ERS used here for simulation of HET is shown in *Figure 3*. The model consists of two layers, one of which is related to the rail embedded in the compound, whereas the other represents a concrete slab on an elastic bed. If, for example, an ERS on a bridge is to be modeled, an extra layer should be added.

The numerical model can analyze various loading cases, ranging from a single load to a whole (high-speed) train [2]. To simulate a Hammer Excitation Test an impulse load has been applied. The load and all essential input parameters

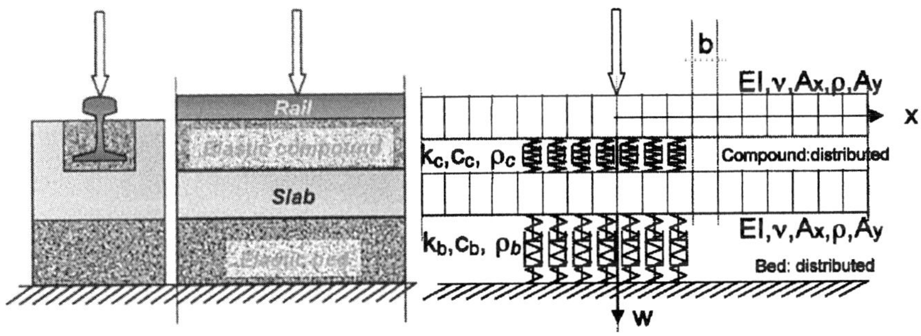

Figure 3. Model of ERS in RAIL (impulse load)

corresponding to this loading case are shown in *Figure 3*. It should be noted that rail surface irregularities (e.g. for analysis of wheel-rail contact forces) can be defined as well. The output data used here comprise the vertical displacements of the rail at the load application point. Using the Fast Fourier Transform, we can calculate the power spectrum ($\widetilde{S}_{XX}$) and cross-spectrum ($\widetilde{S}_{XF}$) of the calculated displacements as well as the power spectrum of the applied impulse force ($\widetilde{S}_{FF}$). The inertance ($\widetilde{H}_{AF}$) and receptance ($\widetilde{H}_{XF}$) FRF are then obtained according to the formulae () and ().

3. TRACK PARAMETER IDENTIFICATION

To ensure that the numerical model adequately describes the behavior of a real track under various loadings one can try to extract some model parameters from the HET measurement data. The FRF, especially the parts near resonant frequencies, supplies detailed information about properties of track components. Here, the model parameters are determined by minimizing the standard deviation σ of the relative differences between the measured H_{AF} and simulated $\widetilde{H}_{AF}$ inertance FRF calculated at certain frequencies. Thus, the parameter identification can be formulated as the following minimization problem:

Minimize

$$D(\boldsymbol{x}) \equiv \underset{i=1,K}{\sigma}\left\{\left|\frac{H_{AF}(f_i,\boldsymbol{x}) - \widetilde{H}_{AF}(f_i,\boldsymbol{x})}{H_{AF}(f_i,\boldsymbol{x})}\right|\right\} \rightarrow \min, \quad \boldsymbol{x} \in R^N \tag{4}$$

where the components of the vector $\boldsymbol{x}$ are the parameters of the model to be identified, and K is the number of frequency discretization points. Here, the minimization problem (4) is solved using (MARS) [5,8]. The method is briefly described below.

3.1 General optimization problem

An optimization problem in a general form can be stated as:
Minimize the objective function

$$F_0(\boldsymbol{x}) \rightarrow \min, \quad \boldsymbol{x} \in R^N \tag{5}$$

subject to the constraints

$$F_j(\boldsymbol{x}) \leq 1, \quad j = 1,...,M \tag{6}$$

and the side constraints

$$A_i \leq x_i \leq B_i, \quad i = 1,...,N \tag{7}$$

where

$\boldsymbol{x} = [x_1,...x_N]^T$ is the vector of design variables;

A_i and B_i are the side limits, which define lower and upper bounds of the i-th design variable.

Components of the vector $\boldsymbol{x}$ represent various parameters of a structure, such as geometry, material, stiffness and damping properties, which can be varied to improve the design performance. Depending on the problem under consideration, the objective and constraint functions (5)-(6) can describe various structural and dynamic response quantities such as weight, reaction forces, stresses, natural frequencies, displacements, velocities, accelerations, etc. The objective function provides a basis for improvement of the design whereas the constraints impose some limitations on behavior characteristics of the structure.

3.2 Approximation concept

The optimization problem (5)-(7) can be solved using a conventional method of mathematical programming. However, for systems with many degrees of freedom the finite element analysis involved in calculation of the objective and constraint functions can be time consuming. As a result, the total computational effort of the optimization might become prohibitive. This difficulty has been mitigated starting from the mid-seventies by introducing approximation concepts [1].

According to the approximation concepts the original functions (5)-(6) are replaced with approximate ones which are computationally less time consuming. Instead of the original optimization problem (5)-(7) a succession of simpler approximated subproblems, similar to the original one and formulated using the approximation functions is to be solved. Each simplified problem then has the following form:

Minimize

$$\tilde{F}_0^k(\boldsymbol{x}) \rightarrow \min, \quad \boldsymbol{x} \in \boldsymbol{R}^N \tag{8}$$

subject to

$$\tilde{F}_j^k(\boldsymbol{x}) \leq 1, \quad j = 1,...,M \tag{9}$$

and

$$A_i^k \leq x_i \leq B_i^k,\ A_i^k \geq A_i,\ B_i^k \leq B_i,\ i = 1,\ldots,N \tag{10}$$

where the superscript k is the number of the iteration step, $\widetilde{F}$ is the approximation of the original function F , A_i^k and B_i^k are *move limits* defining the range of applicability of the approximations.
Since the functions (8)-(9) are chosen to be simple and computationally inexpensive, any conventional method of optimization [3] can be used to solve the problem (8)-(10). The solution of the problem $\boldsymbol{x}_*^k$ is then chosen as starting point for the (k+1)-th step and the optimization problem (8)-(10), re-formulated with the new approximation functions $\widetilde{F}_j^{k+1}(\boldsymbol{x}) \leq 1$, $(j = 0,\ldots,M)$ and move limits A_i^{k+1} and B_i^{k+1}, is to be solved. The process is repeated until the convergence criteria are satisfied.

3.3 MARS optimization technique

Each approximation in (8)-(10) is defined as a function of the design variables x and tuning parameters a (for brevity the indices k and j will be omitted). To determine the components of vector a the following weighted least-squares minimization problem is to be solved [5,8]:

Find vector a that minimizes

$$G(\boldsymbol{a}) = \sum_{p=1}^{P} \{w_p[F(\boldsymbol{x}_p) - \widetilde{F}(\boldsymbol{x}_p,\boldsymbol{a})]^2\}\,. \tag{11}$$

Here $F(\boldsymbol{x}_p)$ is the value of the original function from (5)-(6) evaluated at the point of the design parameters space $\boldsymbol{x}_p$, and P is the total number of such plan points; w_p is a weight factor that characterizes the relative contribution of the information about the original function at the point $\boldsymbol{x}_p$. It should be noted that if the design sensitivities of the original functions are available they can easily be used in the building of approximations [5]. The choice of the approximation function $\widetilde{F}(\boldsymbol{x})$ is important for the success of optimization. In the identification problem presented here a multiplicative model has been used that has the form:

$$\widetilde{F}(\boldsymbol{x}) = a_0 \prod_{i=1}^{P} (x_i)^{a_i}\,. \tag{12}$$

The optimization process is controlled by changing the move limits in each iteration. The iteration process is terminated if the approximations are good, none of the move limits is active and the search subregion is small enough. More information about the approximation models, move limits strategy and most recent developments of the MARS method can be found in [5,9].

4. NUMERICAL RESULTS

This procedure has been applied to determine the material parameters of the compound used in ERS. The components of the vector of the design variables $\boldsymbol{x} = [K_c, C_c, \rho_c]^T$ in the problem () are respectively the stiffness, damping and mass density of the compound. The initial values, lower and upper boundaries of the design variables are collected in *Table 1*. The inertance FRF $\widetilde{H}_{AF}$ in (4) has been obtained using the numerical model implemented in RAIL. The frequency range from 50Hz to 170Hz has been considered since the FRF's in this range are mostly defined by the properties of compound. The MARS method has been used to solve the unconstrained minimization problem (4), and the multiplicative model (12) has been chosen as an approximation of the objective function D in (4). The solution was obtained after 20 iterations. The measured and simulated FRF as well as the

Table 1. Parameters of the identification problem

Parameter	Lower bound	Upper bound	Initial value	Obtained value	Units
K_c	10.0	100.0	16.0	26.3	MN/m/m'
C_c	0.0	50.0	15.0	5.2	kNs/m/m'
ρ_c	0.0	6.0	0.1	2.1	kg/m'

relative difference between the FRF corresponding to the obtained parameters of the compound are shown in *Figure 4*.To estimate the efficiency of the identification procedure, the compound parameters have been obtained using an equivalent single degree of freedom (SDOF) mass-spring system with damping [4]. The mass of the SDOF system representing the rail mass was kept constant, whereas the stiffness and damping were determined from the same measurement data, using a curve-fit procedure of MATLAB. The obtained stiffness and damping parameters of a SDOF system have then been transformed into the stiffness K_c and damping C_c parameters of the RAIL (*Figure 3*). The FRF corresponding to this model and the relative difference are shown in *Figure 5*.

*Figure 4*a and *Figure 5*a show that the MARS model describes the behavior of the real track better than the SDOF model, especially in the neighborhood of the resonant frequency. Also, the overall difference between the measured and simulated FRF is smaller for the MARS model (compare the area under graphs in *Figure 4*b and *Figure 5*b). The same conclusions can be made from *Figure 6* where the results of more accurate calculations with these two models of RAIL are shown. Discrepancy between the measured and simulated FRF (*Figure 4*a) can be explained by insufficient accuracy of the track model. An introduction of non-linear stiffness or/and damping in the numerical model could improve the results.

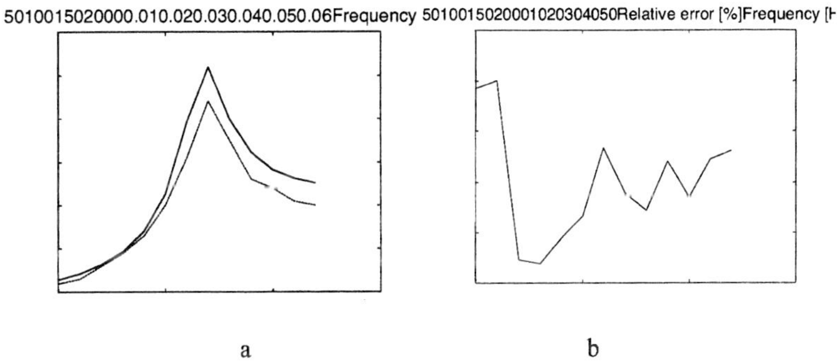

Figure 4. Optimal solution using MARS method: measured (dotted line) and simulated (solid line) inertance FRF (a.); relative difference between measured and simulated FRF (b.)

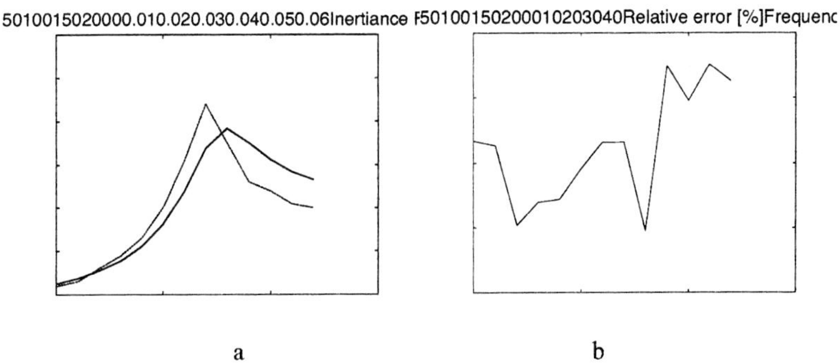

Figure 5. Optimal solution using curve-fit and SDOF mass-spring system: measured (dotted line) and simulated (solid line) inertance FRF (a.); relative difference between measured and simulated FRF (b.)

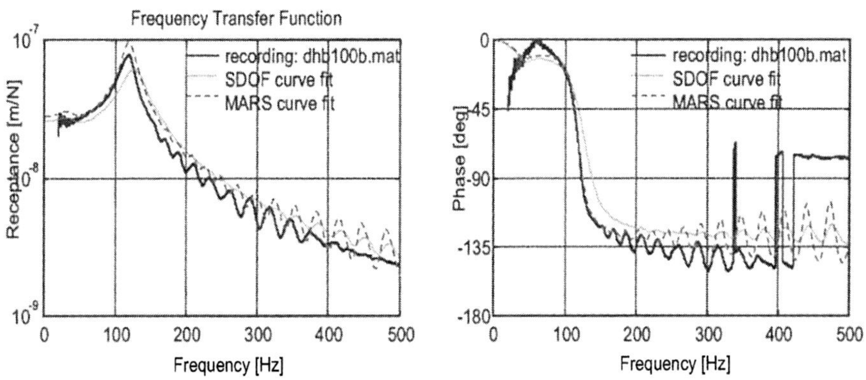

Figure 6. Amplitude and phase of FRF using measured data, MARS and SDOF curve fitting

5. CONCLUSIONS

A procedure for assessment of dynamic properties of components of a railway track has been presented. The procedure includes measuring of the dynamic responses of a track (Hammer Excitation Test), numerical modeling of track behavior (RAIL) and parameter identification. The procedure has been applied to a track without ballast-an Embedded Rail Structure.

Material parameters of elastic compound have been determined based on measurement data using the MARS optimization method. The track model with the obtained parameters of compound can quite adequately describe the behavior of a real track. The results have shown that the MARS method can effectively be used for parameter identification problems.

A better fitting of the measured and simulated data could be obtained by introduction non-linear elements in the numerical model of a track.

References

1. Barthelemy, J.-F.M., Haftka, R.T. (1993) Approximation Concept for Optimum Structural Design – a Review. *Structural Optimization*, **5**, 129-144.
2. Esveld, C., Kok, A.W.M. (1998) Interaction between Moving Vehicles and Railway Track at High Speed. Rail Engineering International, **27**, 3, 14-16.
3. Haftka, R.T. and Gürdal, Z. (1992) *Elements of Structural Optimization.* Third revised and expanded edition, Kluwer Academic Publishers, Dordrecht.
4. De Man, A.P. (1996) Determination of dynamic track properties by means of excitation hammer testing. *Rail Engineering International*, **1996-4**, p.8-9.
5. De Man, A.P. (1999) Interim Report Dynatrack: 2 years of research on the dynamic behaviour of railway track structures (in Dutch), *Report 7-99-111-10*, Railway Engineering Group, TU Delft. ISSN 0169-9288.
6. Markine, V.L. (1999) *Optimization of the Dynamic Behaviour of Mechanical Systems.* PhD Thesis, TU Delft. Shaker Publishing B.V. ISBN 90-423-0069-8.
7. Rasmussen, S., de Man, A.P. (2000) The High Speed Deflectograph and the Hammer Excitation Test: two measurement techniques for determining track performance. *Rail Engineering International*, **2000-1**, p.13-16.
8. Toropov, V.V. (1989) Simulation Approach to Structural Optimization. *Structural Optimization*, **1**, 37-46.
9. Toropov, V.V., van Keulen, F., Markine, V.L., Alvarez, L.F. (1999) Multipoint Approximations Based on Response Surface Fitting: a Summary of Recent Developments. In V.V.Toropov (Ed.) *Proceedings of the Ist ASMO UK/ISSMO Conference on Engineering Design Optimization, Ilkley, West Yorkshire, UK, July 8-9, 1999*, pp.371-381. ISBN 0-86176-650-4

OPTICAL METHOD TO STUDY MATERIAL BEHAVIOUR AT HIGH STRAIN RATES

Jörgen Kajberg
Division of Solid Mechanics, Luleå University of Technology, Luleå, Sweden

Mikael Sjödahl
Division of Experimental Mechanics, Luleå University of Technology, Luleå, Sweden

Keywords: High strain rate, Speckle photography, High-speed photography

Abstract An optical technique to study material behaviour at high strain rates is presented. A small specimen is subjected to a rapid tensional loading in a Hopkinson bar type of arrangement where the loading is measured in the bar following the specimen. A high-speed camera of the image converter type is used to acquire time-resolved images during the loading. These images are evaluated using digital speckle photography to give the in-plane strain field. A calibration procedure is included that compensates for the aberrations in the camera. Experimental results on mild steel are further compared with a finite element calculation with the result that both strain fields show the same general behaviour, but some discrepancies appear that need to be further analysed in the future.

1. INTRODUCTION

Mechanical behaviour at high strain rates is not sufficiently well understood for many engineering materials. Methods to study such phenomena and to determine parameters in constitutive models are therefore of great importance. The purpose of this paper is to present a method, based on digital speckle photography (DSP) [1–3], for the investigation of deformation fields on small specimens subjected to rapid loading. The size of the specimen is chosen small in order to facilitate high strain rates. DSP is a technique that presumes that there exists a characteristic texture (speckle pattern) on the surface, which can be followed when the object deforms using a digital cross-correlation technique. DSP has found great applicability in a number of interesting applica-

P. Ståhle and K.G. Sundin (eds.), IUTAM Symposium on Field Analyses for Determination of Material Parameters - Experimental and Numerical Aspects, 37–49.

tions. Thuvander et.al. [4] used it to study crack tip strain fields in wood at the scale of growth rings, Chiang and Wang studied microscopic deformations using an electron microscope [5], and Synnergren et. al. measured rapid internal deformations using flashed X-rays [6]. In the case of rapid deformations it is advantageous to use a high-speed camera so that the deformation process can be resolved. For this purpose an image converter camera with a CCD-unit is used. Due to the distortion in the image, the speckle displacements from one frame to another cannot be related directly to real displacements. A two step procedure is introduced that compensates for the distortion by using a small calibration object. The results from the experiment on mild steel are compared with a finite element calculation and a static experiment of the same material. The experimental arrangement is described in detail in section 2, while the procedure to evaluate physical strain fields is presented in section 3. The obtained results are presented and discussed in section 4. The paper ends with some concluding remarks about the outcome of the paper.

2. EXPERIMENT

2.1. BASIC IDEA OF THE SET-UP

The basic idea of this set-up is to achieve rapid tensional loading in a small region of a specimen, and measure both the force and the deformation field during the loading. The specimen recieves its rapid loading when it is subjected to impact caused by a projectile. The specimen is mounted on a long slender steel rod (Hopkinson bar) in which the tensional wave from the impact load will propagate. Strain gauges are attached to the rod to facilitate force recording whereas the deformation field is measured with digital speckle photography using a high-speed camera.

The specimen (Figure 1) is a thin mild steel plate with a narrow section in the middle. The narrow section is the interesting part of the specimen because it is where the plastic deformation takes place. The shaded square in Figure 1 shows the field of view.

2.2. SET-UP IN DETAIL

Consider the experimental set-up shown in Figure 2. The projectile is accelerated to a final speed of 6 m/s inside the air gun. On its way the projectile passes two motion detectors mounted to the air gun at known positions. When the front of the projectile reaches each of the motion detectors a signal is sent to an oscilloscope. The oscilloscope gives the time between these two signals and consequently the speed of the projectile. One of the detectors also sends a trigger signal to the four flashlights. The flashlights are activated a short time before the impact in order to achieve full irradiance during the photo session.

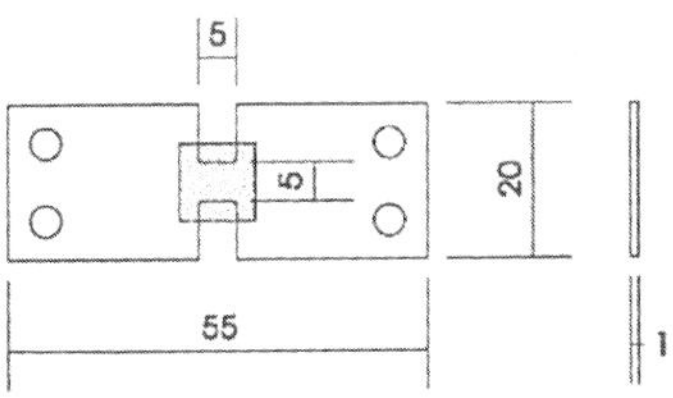

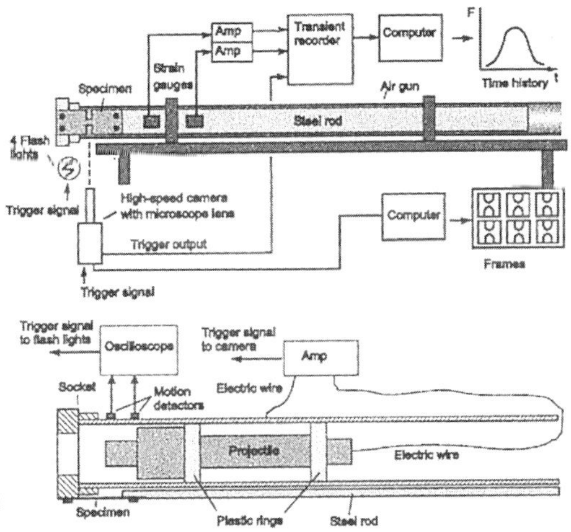

Figure 1 Specimen geometry. The shaded area is the observed region. All numbers are in mm.

Figure 2 Experimental set-up and a cross section of the air gun.

Two plastic rings are mounted on the projectile. These two rings ensure that the projectile fits tight to the air gun. They also work as electric insulators. One electric wire is attached to the end of the projectile and another one to the air gun. The air gun and the projectile form an open electric circuit. When the projectile hits the socket at the muzzle of the air gun a tensional wave begins to propagate through the attached specimen and the steel rod. Furthermore the electric circuit is closed and a trigger signal is transmitted to the camera. The camera immediately sends a signal to the transient recorder, which starts to record the signals from the two strain gauges attached to the steel rod. The data from the transient recorder gives the force history and the elapsed time from impact. The wave propagation speed ($c = \sqrt{E/\rho}$), and the distances between the narrow section of the specimen and the strain gauges are known. Therefore the time when the tension wave reach the specimen can be calculated. A short time after the impact the camera starts its photo session. The delay to the first picture is based on the calculation above and the fact that the time needed for the wave to propagate to the first strain gauge is repeatable with high accuracy.

The synchronisation between the force history and the captured pictures is an important part and it is necessary to get the right force to the corresponding deformation when constitutive parameters shall be determined. In this paper no parametric studies are performed. However, parametric studies are the final goal and therefore the synchronisation is taken into consideration at this early stage.

A high-speed camera has to be used in order to resolve the deformation process. The camera is of the image converter type, ULTRANAC FS 501, with a CCD-unit. The image from the camera consists of up to 4×6 frames. In this paper a configuration of 3×4 frames have been chosen with the advantage that it gives a better spatial resolution for every frame. The sequence of captured frames is shown in Figure 3. The numbers in the frames are the times after the wave has reached the narrow part of the specimen. These are the images used for the strain analysis.

3. EVALUATION OF STRAIN FIELDS

The images in Figure 3 show the deformation of the specimen during the impact. These images give a qualitative idea of the failure mechanisms involved. The purpose of this paper, however, is to use DSP to get quantitative values of the strain field from Figure 3, and hence facilitate direct comparison with a finite element calculation. The main problem with a high-speed camera of the image converter type is that relatively large aberrations are introduced in the imaging. Different parts of the image therefore render the object at slightly different scales. The camera therefore needs to be calibrated so that the measured speckle displacement can be related to physical deformations on the object.

The most fundamental requirement, that there exists a high frequent random pattern at the object surface, has already been fulfilled. The natural texture of the steel surface has been enhanced with diffusely reflecting black and white spray paint (see Figure 3). It is the motion of that random pattern that is analysed with DSP using a digital cross correlation procedure [1]. But there are, however, several other requirements that have to be fulfilled to be able to draw

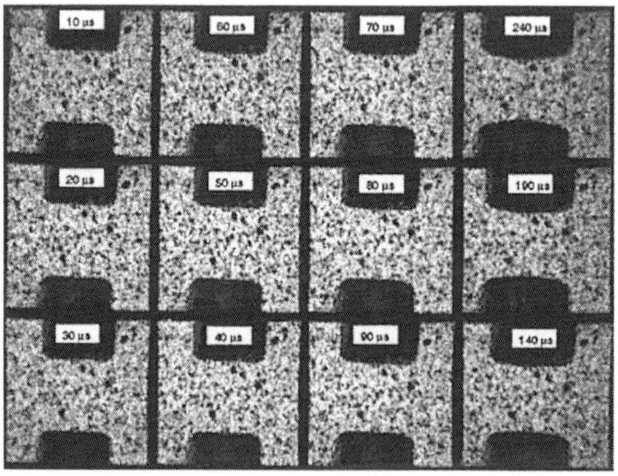

Figure 3 The sequence of captured frames. The times in the frames are the times after the wave has reached the narrow part of the specimen.

quantitative conclusions from Figure 3. First we need to know that the deformations of the same points are followed in all frames. We therefore make an experiment with the object at rest. Everything with the camera and the flash lamps are exactly as they should be during an actual experiment besides that the specimen is not loaded. At this stage we need to decide which resolution to use in the evaluation of the deformation field and which frame to use as a reference. Since the aberrations are expected to be lowest in the middle of the image frame 8 is chosen as reference. A square grid of subimages of size 32×32 pixels separated 8 pixels in the row and column directions, respectively defines the scale in which the strain is evaluated. This resolution is then used throughout the experiments. These subimages are then located in the rest of the frames by applying the DSP algorithm with frame 8 as the reference image. Next the non-linear response to a known deformation in each frame and subimage needs to be established. A calibration plate (placed at the position of the object) is moved ± 200 μm and ± 500 μm, respectively from its initial position. The deformation in each frame and subimage is thereafter evaluated using the DSP algorithm and the response in each subimage is fitted to a third order polynomial in the least squares sense.

The calibration is now complete. Associated with each frame are the positions of the subimages within the frame and the weights of the third order polynomial for each subimage. The physical deformations in Figure 3 can now be evaluated. A set of reference images is acquired with the object at rest prior to the experiment. The deformations in Figure 3 are all calculated relative these reference images and not relative the first frame so that the original grid is maintained throughout the calculations. The result of the DSP calculation is a vector field of in-plane deformations. The result for the 9th frame (90 μs after impact) is shown in Figure 4. In general, however, strains are of more interest than deformations. The in-plane deformation field in Figure 4 in combination with the original grid contains all the information needed to calculate in-plane strains. However, since noise tends to get magnified during numerical differentiation strains need to be calculated with some care. Therefore a first order Savitsky-Golay filter [7] with a filter size of 3×3 points are applied on both components of the deformation field (the local data are fitted to a plane). In-plane strains can then be calculated from the constants in the equation of the planes. With reference to Figure 5 the local variation in the deformation component, u, parallell to the rows of the CCD-detector looks like

$$u = u_0 + ax + by \Leftrightarrow u = u_0 + \frac{\partial u}{\partial x}x + \frac{\partial u}{\partial y}y, \quad u_0 = const,$$

where the strain component along the detector rows is given by $\frac{\partial u}{\partial x}$. In order to describe large deformations, the logarithmic strain definition is used. The

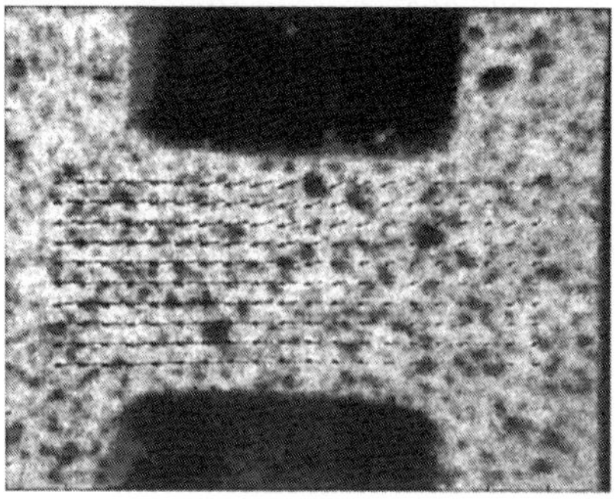

Figure 4 Measured displacement field 90 μs after impact.

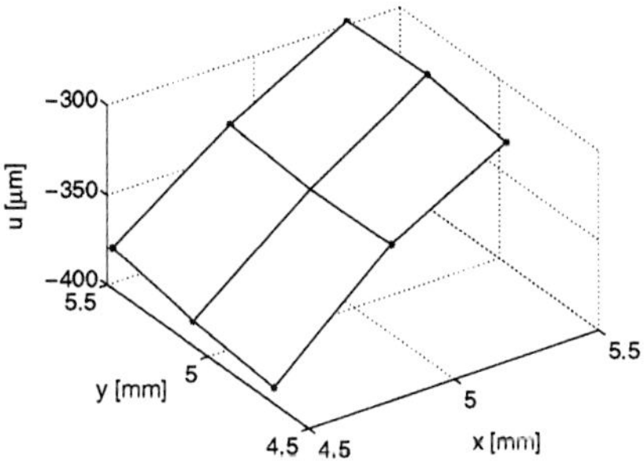

Figure 5 The speckle displacement component parallel to the loading direction for 9 points in the middle of the observed area.

strain component ε_x, then finally has the form

$$\varepsilon_x = \ln\left(1 + \frac{\partial u}{\partial x}\right),$$

which is the strain component presented in this paper.

4. RESULTS AND DISCUSSION

The recorded force from the two strain gauges show good agreement, but the right one in Figure 2 has a less noisy signal. Its result is presented in

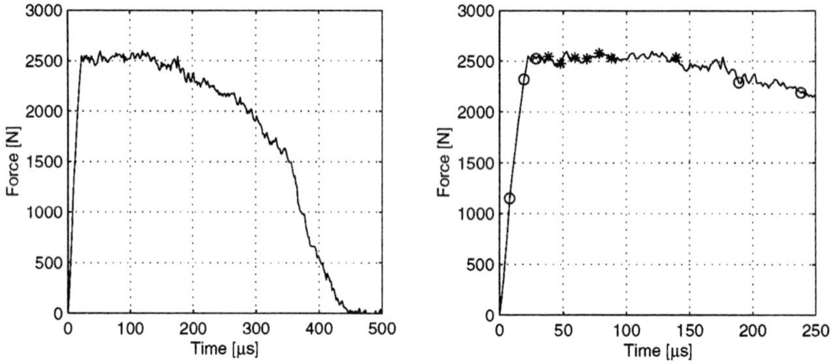

Figure 6 Force history. The right picture shows the part of the force history used in the FE-simulation. Stars and circles indicate the times when the photographs have been taken.

Figure 6, where the left picture shows the history until failure and the right one shows the force for the first 250 μs. The times for the calculated strain fields are marked with stars and the circles indicate the frames where no strains are determined, due to too small or too large relative speckle displacements. The outgoing wave is assumed to the same as the incoming tensional force. This assumption is only valid if the influence of inertia on the stress distribution in the narrow part of the specimen is assumed to be small. Further, possible distortion of the tensional wave caused by the attachment of the specimen to the steel rod has not been considered. The negligence of the inertia and the influence of the attachment need to be studied more in detail.

The measured logarithmic strain fields in the evaluated regions for 7 different times between 40 μs and 140 μs are shown in Figure 7. Note that for all times except the 140 μs time instant, the same strain interval, 0 – 0.1, is used. The interval is chosen to emphasise the development of the strain. The resulting strain field from the last time instant has, however, higher values and the interval is therefore different. As a preliminary comparison the results from a FE-calculation is also shown in Figure 7. Because of the high strain rate a rate dependent constitutive model has been used. See Appendix for more details. The resulting logarithmic strain can be studied in Figure 7 where the same scaling as for the experimental results have been used.

Several interesting observations emerge from the comparison between the experimental and numerical results. Both results show the same general appearance that the highest strain is located in the middle of the specimen and that a cross shaped strain region appears from the corners of the narrow region. In the FE-pictures, however, it is very easy to see the development of the strain, and how the plastic region becomes larger. This increase in the plastic region is not as profound in the experimental strain fields. There might be several reasons for this discrepancy between the experimental and the numerical results. One important difference is that the spatial resolution in the experimental results is limited by the size of the subimage used to evaluate the deformation. The subimage also tends to smear out the deformation field, which results in a somewhat blurred strain field when differentiated. The mesh on the other hand gives the spatial resolution in the FE-calculation. The difference in spatial resolution can, however, not explain the whole difference in the two approaches so that other causes have to be found in the numerical model. Matching the model with the experimental results will be one of the interesting challenges in the future.

Another interesting quantity to evaluate is the strain rate experienced by the material. Therefore the mean strain for the region with the highest strain has been calculated and plotted against time. The region is marked with a rectangle in Figure 7g. The development of the mean strain is presented in the left graph in Figure 8 together with a fitted third order polynomial. The right graph shows

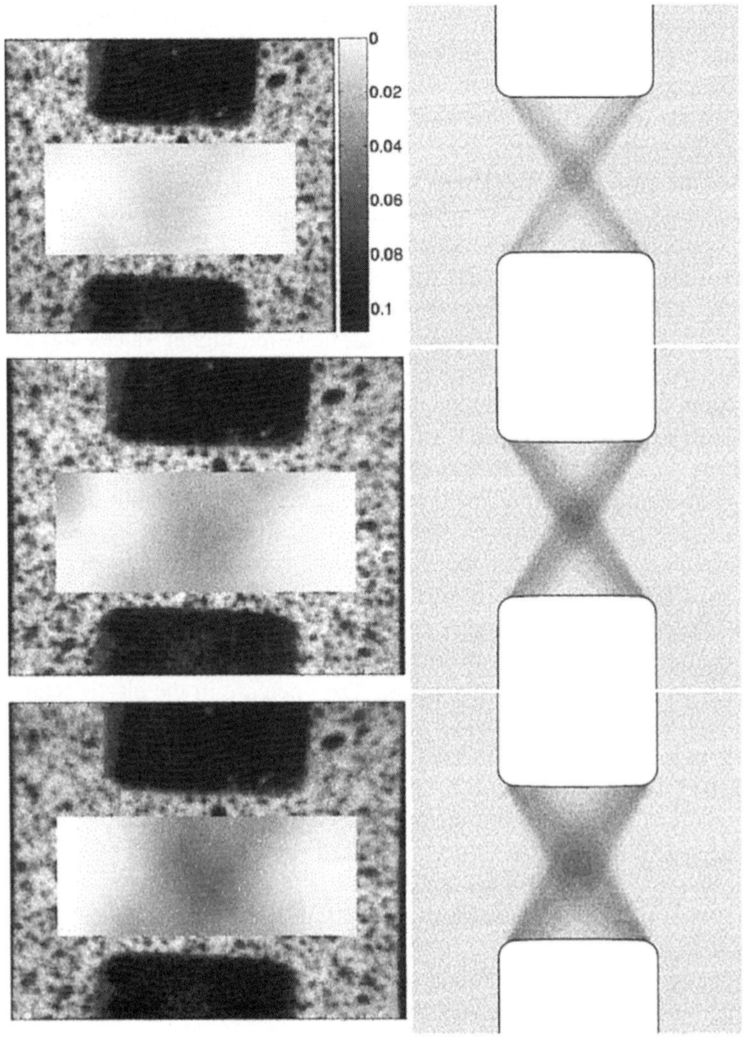

Figure 7 Measured and FE-calculated strain fields, 40, 50 and 60 μs after impact.

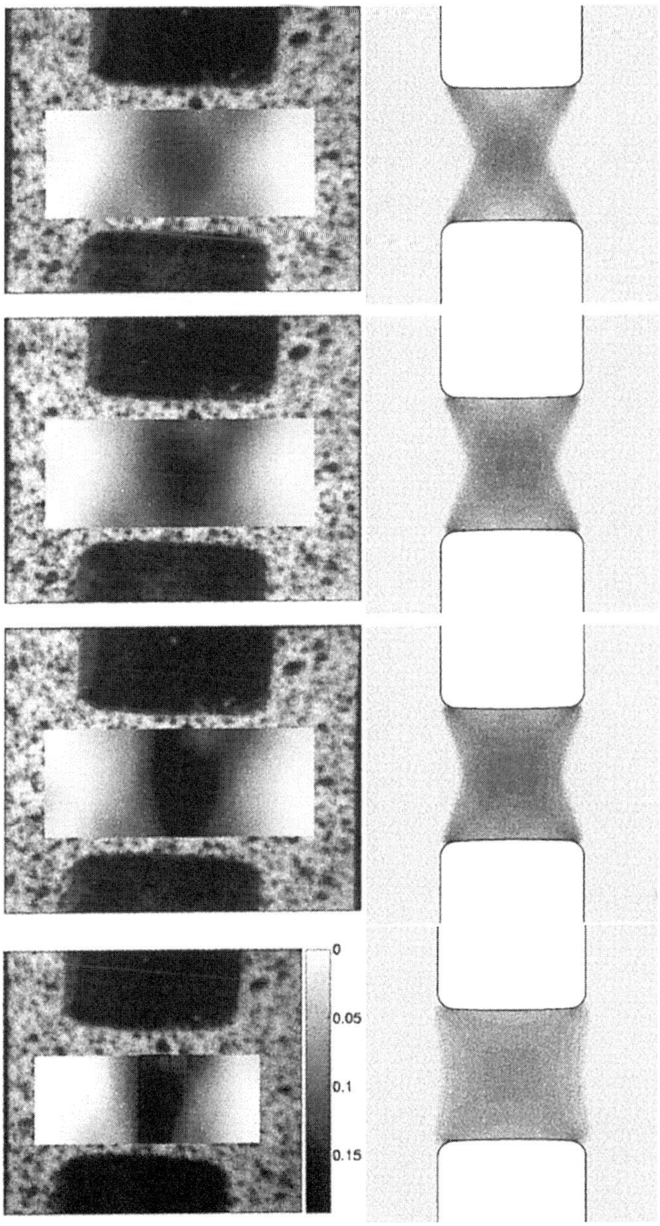

Figure 7 (continued)
Measured and FE-calculated strain fields, 70, 80, 90 and 140 μs after impact.

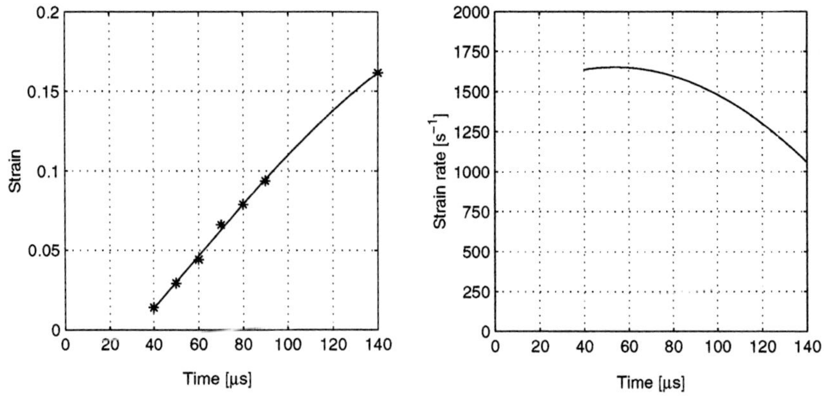

Figure 8 Strain and strain rate vs. time

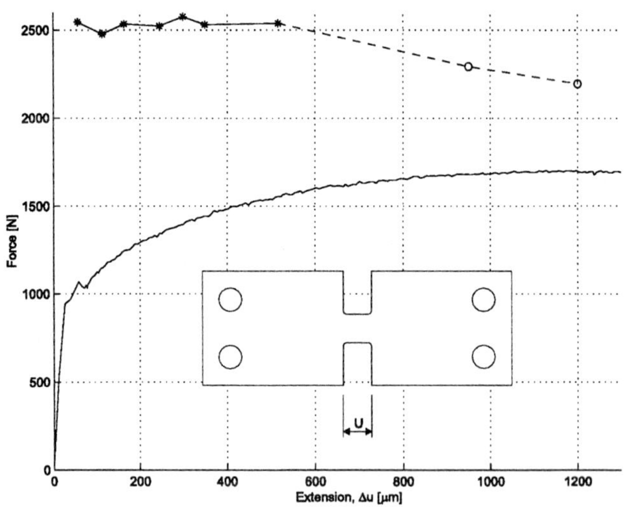

Figure 9 Force vs. extension for the static and dynamic test

the strain rate evaluated as the time derivative of this polynomial. As shown in the figure, the maximum strain rate is estimated to be about 1650 s^{-1} and somewhat lower at longer times.

A last interesting question is, if it is possible to detect any effects of the expected viscoplastic hardening due to the high strain rates presented in Figure 8. One effect should be that the force when yield occurs becomes higher than the required force for a static condition. In order to answer that question a static tension test was performed. The result of this test is shown in Figure 9. The force vs. extension from the dynamic test is presented in the same figure. The extension, Δu (the definition is shown in the figure), was calculated by taking the average speckle displacement at the left boundary of the narrow section minus the corresponding displacement at the right boundary. The last two values, marked with a circle, are extensions estimated by counting the increase of pixels between the boundaries mentioned above. Therefore the two last extensions have less reliability. In this case the force when yield occur has increased with a factor about 2.5, and this is in good agreement with Campbell and Ferguson [8], who propose an increase about 2.3 for mild steel at this strain rate.

5. CONCLUSIONS

An optical method designed to study material behaviour at high strain rates has been presented. The load was applied in a tensional Hopkinson bar arrangement where the outgoing wave was recorded at two positions on the bar using electrical strain gauges. The resulting strain field in the object was measured using an image converter type high-speed camera combined with digital speckle photography (DSP). A calibration procedure is included that compensates for the aberrations in the camera and makes quantitative evaluation of the in-plane strain field possible. The results from the DSP calculation have been used in three different ways. Firstly, the logarithmic strain field component parallel to the loading direction was calculated and compared with a numerical simulation. These two approaches showed the same general material behaviour, but some discrepancies were clearly visible the cause of which needs to be explained in the future. Secondly, the mean strain evolution within the most strained region have been calculated and plotted against time to facilitate strain rate calculation. The maximum strain rate was calculated to be 1650 s^{-1} which corresponds well with the expected strain rate. The third application of the DSP results was to investigate if viscoplastic hardening could be detected. The differences in displacement between the two sides of the sample was therefore calculated and compared with the extensiometer readout from a static experiment using the same material. Using this technique an increase in the yield level of about 2.5 was estimated which compares well with previously tabulated values.

Acknowledgments

The authors are grateful to Per Synnergren, Jan Granström and K-G Sundin for their help during the experiments and preparation of this paper.

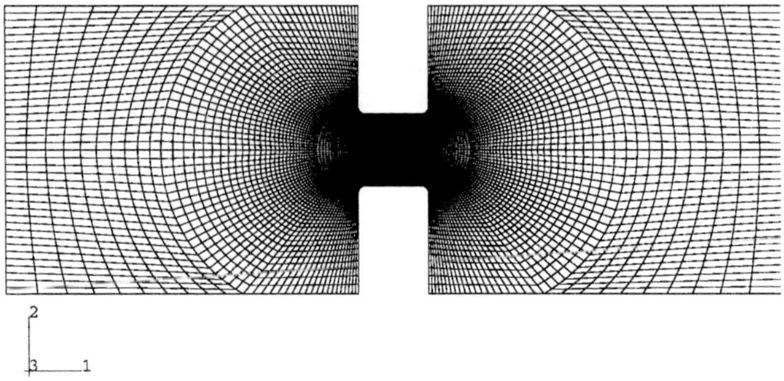

Figure 1 The mesh

Appendix: FE-simulation

In order to show a preliminary comparison between the experimental and the numerical results a FE-calculation has been performed. The FE-code used for the calculation is ABAQUS/EXPLICIT 5.8 [9]. Because of the high strain rate, a viscoplastic constitutive model has been used. The constitutive model used in this simulation is the one suggested by Perzyna[10], in which the viscoplastic strain rates are given by

$$\dot{\varepsilon}_{ij}^{p} = \begin{cases} \frac{3}{2}\gamma\left(\frac{\sigma_e}{\sigma_Y(\varepsilon_p)} - 1\right)^n \frac{S_{ij}}{\sigma_e}, & \sigma_e > \sigma_Y(\varepsilon_p) \\ 0, & \sigma_e \leq \sigma_Y(\varepsilon_p) \end{cases}$$

where γ is the strain rate sensitivity, n is a strain rate exponent, σ_Y is the yield stress, σ_e is the effective stress according to von Mises and S_{ij} is the deviatoric stress. The two unknown parameters, γ and n, are chosen to 25000 s^{-1} and 7, respectively. Further plane stress has been assumed in the specimen. The applied tensional force is shown in Figure 6. Figure 1 in the appendix presents the whole mesh.

References

[1] Sjödahl, M. Electronic speckle photography: increased accuracy by non-integral pixel shifting. *Appl. Opt.*, Vol. 33, pp. 6667-6673, 1994.

[2] Sjödahl, M. Accuracy in electronic speckle photography. *Appl. Opt.*, Vol. 36, pp. 2875-2885, 1997.

[3] Synnergren, P. and Sjödahl, M. A stereoscopic digital speckle photography system for 3-D displacement field measurements. *Opt. and Las. in Eng.*, Vol. 31, 425-443, 1999.

[4] Thuvander, F., Sjödahl, M. and Berglund, L. A. Measurements of crack tip strain field in wood at the scale of growth rings. To appear in *J. Mat. Sci.*, 2000.

[5] Chiang,F. P. and Wang, Q. Speckle interferometry with nanoparticles. *Proc. SPIE*, Vol. 3407, pp. 310-317, 1998.

[6] Synnergren, P., Goldrein, H. T. and Proud, W. G. Application of Digital Speckle Photography to Flash X-ray Studies of Internal Deformation Fields in Impact Experiments. *Appl. Opt.*, Vol. 38, pp. 4030-4036, 1999.

[7] Press, Teukolsky, Vetterling, Flannery. *Numerical recipies in C, 2nd ed.*, (Cambridge University Press, Cambridge UK, 1992).

[8] Campbell, J. D. and Ferguson, W. G. The temperature and strain rate dependence of the shear strength of mild steel. *Phil. Mag.*, Vol. 21, 63-82, 1970.

[9] ABAQUS User's manual, version 5.8, Hibbit, Karlsson and Sorensen, Inc, 1998.

[10] Perzyna, P. The Constitutive Equations for Rate-Sensitive Plastic Materials. *Q. Appl. Math.*, Vol. 20, 321-332, 1963.

DETERMINATION OF CYLINDRICAL ELASTIC ORTHOTROPY IN WOOD BY USE OF DIGITAL SPECKLE PHOTOGRAPHY

Lars Olof Jernkvist
Quantum Technologies AB, Uppsala Science Park, SE-751 83 Uppsala, Sweden.

Fredrik Thuvander
Karlstad University, SE-651 88 Karlstad, Sweden

Abstract. The orthotropic elastic properties in the radial-tangential plane of Norway spruce were determined by use of an optical deformation measurement technique and inverse finite element modelling. White light digital speckle photography in combination with optical microscopy was used for measuring the local deformations in a single growth ring of the material. As a result of the high spatial resolution in the measurement, it was possible to determine the radial variation of the in-plane coefficients of elasticity across the growth ring; the early springwood and the late summerwood were found to have very different stiffness properties.

1. INTRODUCTION

Current finite element software permits the mechanical behaviour of materials with heterogeneous internal structure to be studied on a micromechanical basis; from knowledge of the geometrical configuration and mechanical properties of individual constituents, the complex internal anatomy can be modelled, and the overall performance of the material investigated. Such analyses are frequently used in *e.g.* composite science or biomechanics, usually with a view to gain understanding of processes leading to damage or failure in the materials [1].

However, micromechanical analyses are of small value, unless the mechanical properties of the individual constituents are well established. This is rarely the case, since quantitative characterisation of microscopic building blocks, such as fibres or biological cells, is troublesome. The characterisation can seldom be performed by use of conventional testing techniques, and there is a need for experimental methods that lend themselves to small and fragile samples. Several optical measuring techniques, such as holographic interferometry or speckle photography, answer to these requirements [2]. These are highly accurate, non-contact methods. Moreover, they provide information about the entire deformation pattern of a specimen surface, and are therefore well suited for measurements on objects undergoing non-trivial modes of deformation, caused for instance by multiaxial loading or material heterogeneity. Speckle

P. Ståhle and K.G. Sundin (eds.), IUTAM Symposium on Field Analyses for Determination of Material Parameters - Experimental and Numerical Aspects, 51–61.

photography is particularly useful for micromechanical materials testing, since the technique is scalable and can be combined with microscopy for deformation measurements on very small objects. In white light speckle photography, the in-plane deformation of a strained object is obtained from the displacement of a speckle pattern attached to the object's surface. The relative displacement of the speckle pattern is calculated by comparing two images, one taken before and one after loading the object, on a subimage basis. Provided that the speckle pattern is clear and random, corresponding subimages before and after loading can be identified with a pattern recognition algorithm, and the relative displacement of each subimage can be calculated. The images of the speckle pattern can be recorded either on photographic film or by use of an ordinary off-the-shelf CCD camera. In the latter approach, which is usually called digital speckle photography (DSP), the digitally recorded images can be processed by the pattern recognition algorithm almost in real time. The DSP-method is thus both fast and inexpensive, and has therefore found a wide range of applications in experimental mechanics [3].

In this paper, DSP has been used for determining the variation in elastic properties of spruce wood on a microstructural level. Wood has a complex heterogeneous internal structure, and the mechanical properties of the material are strongly influenced by the anatomy [4]. In spruce and other softwoods, about 95% of the volume is occupied by hollow elongated cells, called fibres or tracheids, which are aligned with the longitudinal direction of the tree trunk; see figure 1. The remaining volume comprises longitudinal resin channels and radially aligned ray cells. The length of the tracheids is approximately 3 mm, and the diameter ranges from 10 to 30 μm. The diameter, as well as the cell wall thickness, vary over each annual growth ring: wide and thin-walled cells are formed in the spring, whereas narrow and thick-walled cells are formed later in the season of growth. The annual alteration between these two kinds of material, termed earlywood and latewood, produces a repetitive variation in both density and stiffness along the radial direction of the trunk.

From a macroscopic point of view, wood exhibits cylindrical orthotropic properties as a result of the growth ring formation and the longitudinal alignment of the tracheids. As shown in figure 1, the principal axes of orthotropy are approximately given by the radial, tangential and longitudinal directions (R,T,L). Under plane stress conditions, the orthotropic relation between stress and strain in the (R,T)-plane can be written

$$\begin{pmatrix} \sigma_R \\ \sigma_T \\ \tau_{TR} \end{pmatrix} = \frac{1}{\left(1-\nu_{RT}\nu_{TR}\right)} \begin{bmatrix} E_R & E_R\nu_{TR} & 0 \\ E_T\nu_{RT} & E_T & 0 \\ 0 & 0 & G_{TR}\left(1-\nu_{RT}\nu_{TR}\right) \end{bmatrix} \begin{pmatrix} \varepsilon_R \\ \varepsilon_T \\ \gamma_{TR} \end{pmatrix}, \quad (1)$$

where the stiffness matrix is symmetric, *i.e.* $E_T\nu_{RT} = E_R\nu_{TR}$. In computational analyses, the components of the stiffness matrix are usually considered to be constant, *i.e.* the spatial variability of the stiffness is neglected, and averaged or homogenised properties are used. These homogenised material properties are representative of the real material, as long as the stress or strain state imposed by the loading is tolerably uniform over length scales, exceeding the growth ring width. Unfortunately, this condition is often violated close to notches and cracks in the material, and the variation in stiffness across growth rings must therefore sometimes be considered in fracture mechanics analyses of wood [5]. The radial stiffness variation across growth rings is however unknown; it has so far only been possible to estimate it from the difference in tracheid geometry between earlywood and latewood [6]. In order to verify and validate this kind of estimations, the radial variation of the elastic parameters E_R,

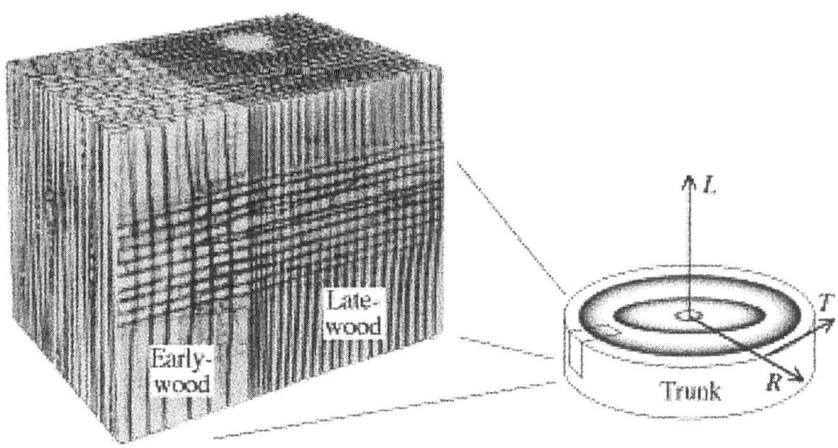

Fig 1. The anatomy of a typical softwood. Microphotograph from [4].

E_T, G_{TR} and ν_{TR} across a growth ring in spruce was experimentally determined by use of DSP. The work is presented in the sequel, with special emphasis given to the numerical methods applied for evaluating the material properties from the measured deformation fields.

2. EXPERIMENTAL PROCEDURE

2.1 Material and specimen preparation

A flat, dogbone-shaped tensile specimen as shown in figure 2 was fabricated from a board of Norway spruce (*Picea abies*), that had been kiln dried in a standard commercial process. The moisture content of the material was 12.2 weight-%, and the average density 447 kgm^{-3}. The specimen was cut out with its main axes aligned with the principal axes of material orthotropy. The thickness was 3.28 mm, and the dimensions in the radial-tangential plane were 25.0x13.0 mm. The growth ring under study was located at the chamfered mid-position of the sample (shaded in figure 2), where the tangential width was reduced from 13.0 to 5.40 mm.

2.2 Test procedure

A tensile load of 3.0 MPa was applied in the radial direction through a miniature tensile stage, (Minimat, Ploymer Labs), and the resulting deformations were studied on a flat surface in the radial-tangential plane of the growth ring by use of an optical microscope, cf. figure 2. A CCD camera was mounted to the microscope, and images were recorded on a PC via a frame grabber with a resolution of 576x768 pixels. The magnification obtained through this arrangement was 2.10μm/pixel, and 10 images were therefore needed to capture the entire growth ring. With the intention of simplifying the subsequent image analysis, these partly overlapping images were compiled into a full image of the growth ring, comprising 1265x2664 pixels.

2.3 Displacement field measurement

The in-plane deformation of the loaded specimen was determined by use of a DSP-algorithm, in which cross-correlation of the images recorded before and after loading is performed in the frequency domain via 2D fast Fourier transformation [7]. The algorithm allows displacement measurements with standard deviations as low as 1% of the pixel size, but to reach this accuracy, a high-contrast speckle pattern with a mean speckle size corresponding to approximately two pixels is needed [8]. To meet this requirement, the specimen surface was dusted with Ø 5 μm carbon toner particles from a laser printer, which gave a random speckle pattern with the desired frequency content and contrast.

The accuracy and spatial resolution in the displacement measurement are dependent on the size of the subimages that are used for cross-correlating the speckle patterns in the deformed and undeformed state. The use of large subimages is desirable, since it leads to reliable pattern recognition, and thus to smaller errors. On the other hand, small subimages are needed in order to obtain a high spatial resolution in the measurements. Hence, the optimum subimage size is a balance between accuracy and resolution. In our experiment, the pattern recognition algorithm was applied to subimages of size 32 by 64 pixels in the radial and tangential directions, respectively. The displacements were evaluated in 38x40 equidistant points across the growth ring, which corresponds to a spatial resolution of 67.2 μm in the radial and 134.4 μm in the tangential direction.

3. EVALUATION OF DISPLACEMENT FIELDS

3.1 Determination of in-plane strain components

The strain components in the radial-tangential plane were obtained from numerical differentiation between adjacent points in the measured displacement fields. To this end, the measured displacement fields were first filtered, in order to reduce the impact of random errors on the differentiation. The filtering can be described as a neighbourhood operation, performed on the measured displacements in each discrete point (R_i, T_j) of the displacement field. The filtered displacement u' was calculated through

$$u'(R_i,T_j) = \sum_m \sum_n u(R_m,T_n)\, h(i-m, j-n) \; . \tag{2}$$

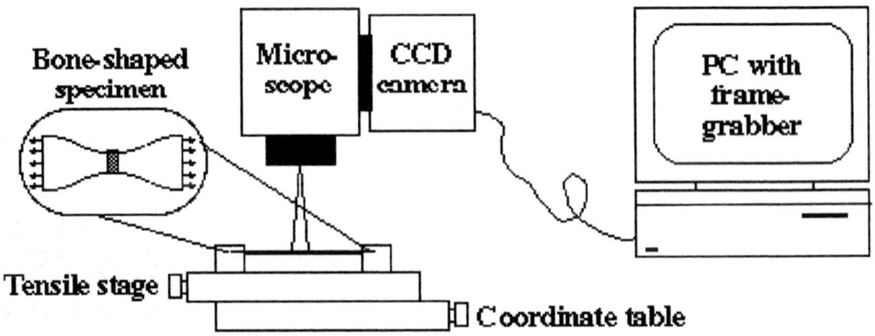

Fig 2. Experimental set-up and apparatus.

Here, u is the measured data and the discrete filter kernel h is given by

$$h(i,j) = \begin{cases} 1/2 & \text{for } i = j = 0 \\ 1/16 & \text{for } |i| = 1 \text{ and } |j| \leq 1 \\ 1/16 & \text{for } |j| = 1 \text{ and } |i| \leq 1 \\ 0 & \text{otherwise} \end{cases} . \qquad (3)$$

A contour plot of the radial strain field is shown in figure 3, together with a micrograph of the investigated growth ring. A coordinate system (r,t) is also introduced, the axes of which are normalised with the growth ring width, w = 2.44 mm. Although the strain field is impaired by noise and local variations in the cellular structure, it is evident that there is a large-scale variation in strain, not only in the radial, but also in the tangential direction. The radial variation is caused by the gradual change in stiffness from earlywood to latewood across the growth ring, whereas the tangential variation is due to the abrupt change in stiffness across the latewood-earlywood interface at $r = 0$. As can be seen in figure 3, this discontinuity also induces a Poisson effect, *i.e.* an apparent stiffening of the soft earlywood close to the interface. In the central part of the specimen, where the Poisson effect is most pronounced, the maximum strain amplitude is found at a significant distance from the interface. Similar results were obtained also for the tangential strain component and the in-plane shear. In this context, it should be noticed that shear was caused by the abrupt change in stiffness across the latewood-earlywood interface, and not by the external loading.

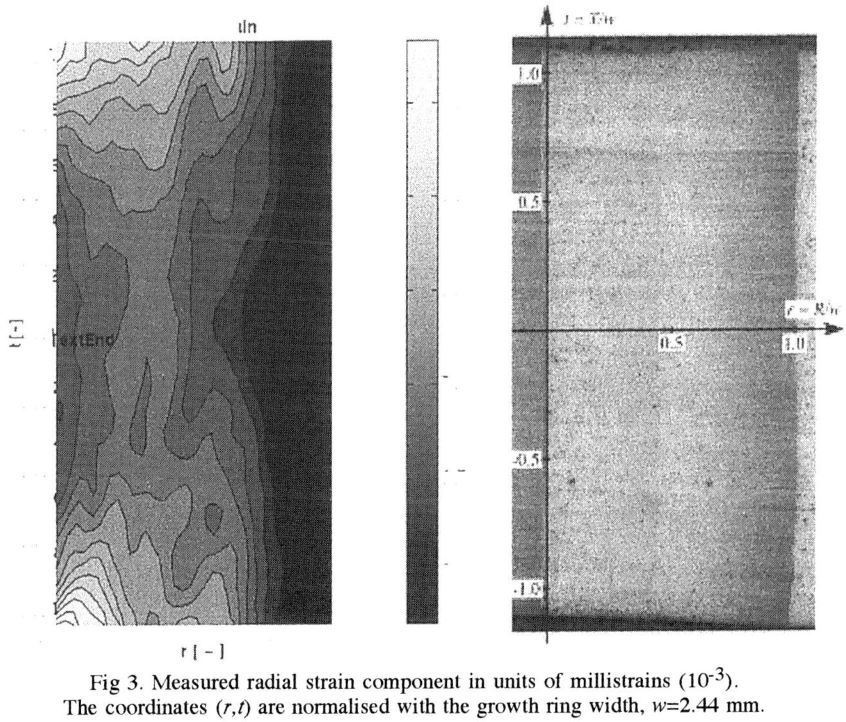

Fig 3. Measured radial strain component in units of millistrains (10^{-3}).
The coordinates (r,t) are normalised with the growth ring width, w=2.44 mm.

3.2 Determination of in-plane stiffness properties

The coefficients of elasticity in eq. (1) were determined from the measured strain fields by inverse modelling. A plane stress finite element model of the growth ring was used, in which measured displacements at the boundaries $r = 0$ and $r = 1$ were applied as boundary conditions. Departing from assumed stiffness properties, the in-plane strains were calculated and compared with their measured counterparts. By repeating this procedure, the assumed stiffness properties were iteratively improved with respect to best possible agreement between calculated and measured strain fields. In order to make the problem tractable, it was assumed that the four independent coefficients of elasticity in eq. (1) vary only with the radial position in the growth ring. This assumption is vindicated by the cellular structure, which exhibits a systematic variation only in the radial direction. Hence, the applied constitutive law could in vectorial form be written $\bar{\sigma}(r,t) = \bar{\boldsymbol{D}}(r)\cdot\bar{\varepsilon}(r,t)$.

Before starting the finite element inverse modelling, the number of unknown stiffness parameters was reduced by exploiting the equations of equilibrium in combination with the measured strain fields. In the local (r,t)-coordinate system, the growth ring curvature can be neglected, and the in-plane equations of equilibrium may be written

$$\frac{\partial \sigma_R(r,t)}{\partial r} + \frac{\partial \tau_{RT}(r,t)}{\partial t} = 0 \quad \text{and} \tag{4}$$

$$\frac{\partial \sigma_T(r,t)}{\partial t} + \frac{\partial \tau_{RT}(r,t)}{\partial r} = 0\,, \tag{5}$$

where $r \in [\,0,1\,]$ and $t \in [-1.10,\ 1.10\,]$. By combining eq. (1) with eq. (4), and integrating over $0 \le r \le r_i$, $-1.10 \le t \le 1.10$, we find

$$\frac{1}{2.20}\int_{-1.10}^{1.10}\sigma_R(r_i,t)dt = \frac{D_{11}(r_i)}{2.20}\int_{-1.10}^{1.10}\varepsilon_R(r_i,t)dt + \frac{D_{12}(r_i)}{2.20}\int_{-1.10}^{1.10}\varepsilon_T(r_i,t)\,dt = 3.0\cdot 10^6, \tag{6}$$

which simply states that the average radial stress is 3.0 MPa for all radial positions r_i. If r_i are taken as the radial positions at which the measured strain fields are sampled, eq. (6) allows $D_{11}(r_i)$ to be expressed in terms of $D_{12}(r_i)$ by use of the measured strain data.

A similar relation between $D_{22}(r_i)$ and $D_{21}(r_i)$ is provided by the condition that the tangential stress is zero at the specimen free boundaries ($t = \pm 1.10$);

$$\sigma_T(r_i, t = \pm 1.10) = D_{21}(r_i)\,\varepsilon_R(r_i, t = \pm 1.10) + D_{22}(r_i)\,\varepsilon_T(r_i, t = \pm 1.10) = 0\,. \tag{7}$$

By use of eqs. (6) and (7), $D_{11}(r)$ and $D_{22}(r)$ are eliminated as independent stiffness parameters, and the inverse modelling problem is thus reduced to finding the radial variation of D_{33} and D_{21} across the growth ring. This can be done through eq. (5), which need be commented: In a homogeneous material under the current conditions of uniaxial loading, this equation would be trivially satisfied and thus provide no additional information for determining the material stiffness properties. In the growth ring, however, the abrupt change in stiffness between latewood and earlywood induces a complex biaxial variation in stress and strain. The derivatives in eq. (5) are therefore non-zero, and the equation thus provides the information needed to determine the radial variation of D_{33} and D_{21}.

Consider integration of eq. (5) over a rectangular area in the growth ring, $r_{i-1} \leq r \leq r_{i+1}$ and $t_{j-n} \leq t \leq t_{j+n}$, where $n \geq 1$. The area comprises 3x(2n+1) sampling points of the strain fields. Substitution of eq. (1) into eq. (5), followed by integration yields a system of linear equations for D_{21}, D_{22} and D_{33}:

$$D_{21}(r_i)\left[\varepsilon_R(r_i,t_{j+n})-\varepsilon_R(r_i,t_{j-n})\right] + D_{22}(r_i)\left[\varepsilon_T(r_i,t_{j+n})-\varepsilon_T(r_i,t_{j-n})\right] +$$
$$\frac{n\Delta t}{(2n+1)\Delta r} D_{33}(r_i)\sum_{k=j-n}^{j+n}\left[\gamma_{RT}(r_{i+1},t_k)-\gamma_{RT}(r_{i-1},t_k)\right] + \qquad (8)$$
$$\frac{n\Delta t}{(2n+1)\Delta r}\left[D_{33}(r_{i+1})-D_{33}(r_{i-1})\right]\sum_{k=j-n}^{j+n}\gamma_{RT}(r_i,t_k) = 0 \;.$$

Here, (r_i, t_j) are discrete sampling points for the strain fields, whereas Δr and Δt are the radial and tangential distances between adjacent points. By use of eq. (7), $D_{22}(r_i)$ can be expressed in terms of $D_{21}(r_i)$ and directly substituted into eq. (8). The resulting relation between $D_{21}(r_i)$ and $D_{33}(r_i)$ should be valid for arbitrary areas of integration, *i.e.* for any values of j and n. By selecting a number of different combinations of j and n, an overdetermined system of linear equations for $D_{21}(r_i)$ and $D_{33}(r_i)$ is obtained. The system can be solved in a least-square sense after normalisation; at least one degree of freedom must be fixed, in order to obtain non-trivial solutions to eq.(8). Hence, by letting for instance $D_{21}(r_1) = C_0$, all other stiffness parameters for this particular value of C_0 can be found through eqs. (6)-(8). The inverse finite element modelling is thereby reduced to an iterative procedure for finding the value of a single parameter C_0, which results in the best possible agreement between calculated and measured strain fields. The "best possible agreement" was found by minimising the cost function

$$f(C_0) = \sum_{i=1}^{37}\sum_{j=1}^{39}\left|\frac{\varepsilon_R^{calc}(r_i,t_j,C_0)}{\varepsilon_R^{meas}(r_i,t_j)} - 1\right| . \qquad (9)$$

Here, the summation is taken over the relative differences between measured and calculated radial strain in each discrete point (r_i, t_j). The minimum to $f(C_0)$ in eq. (9) was found by a full search over C_0 in the range of interest. The results in terms of the radial variations for E_R, E_T, G_{TR} and ν_{TR} are shown in figure 4.

4. DISCUSSION

It is evidenced by figure 4, that there are significant differences in stiffness properties between earlywood and latewood. The differences are largest for E_T and G_{TR}, whereas E_R and ν_{TR} show a more moderate variation across the growth ring. These results are further discussed and compared with predictions from micromechanical models of the wood anatomy in [9]. Here, the discussion will be confined to the accuracy of the presented results.

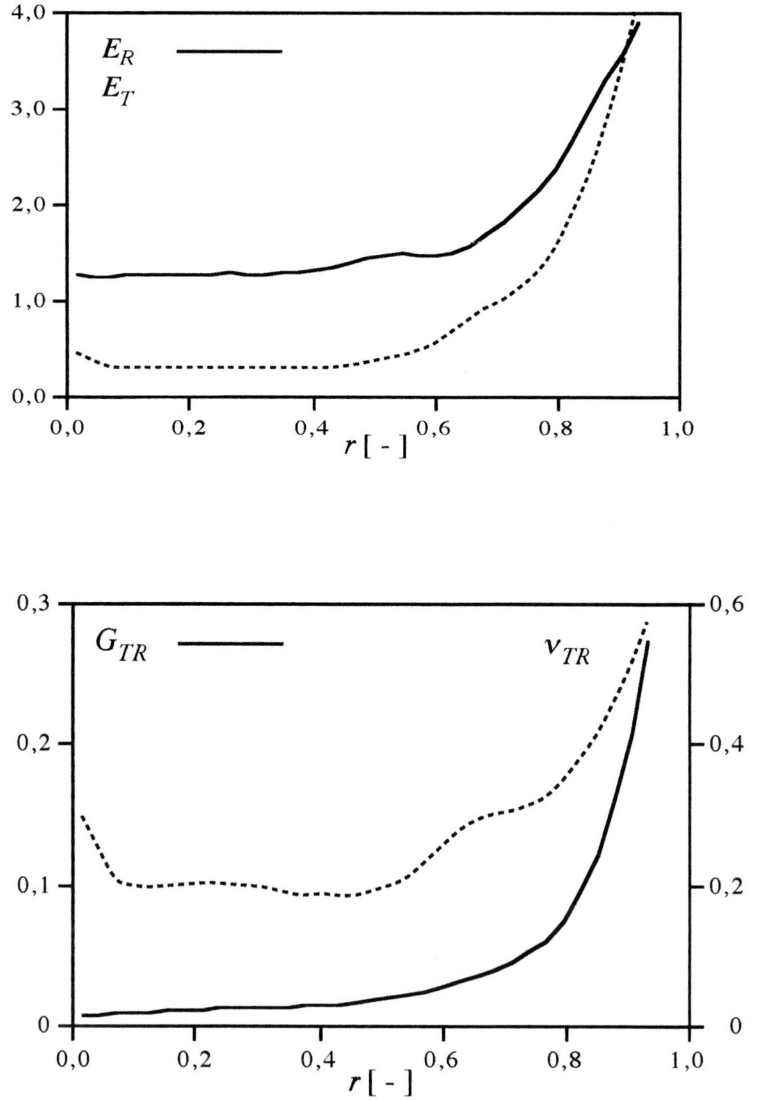

Fig 4. Measured coefficients of elasticity with respect to radial position in the growth ring.

4.1 Measurements of displacements and strains

For the purpose of avoiding in-elastic deformation of the material, the strain amplitudes were kept low in the experiment. Since it was suspected that the measured strains could be corrupted with noise, an additional deformation measurement was performed in order to determine the typical magnitude of random errors. The specimen, the test set-up and methodology were exactly the same as before, but the specimen was subjected to a rigid body translation instead of a radial tensile stress. The standard deviations of the measured radial and tangential displacements under the rigid body translation were 2.5 and 4.4% of the pixel size respectively. The random errors in our investigation are thus somewhat larger than the theoretical limit of accuracy, 1%, which has been reported for the applied DSP-algorithm [7]. Since a rigid body translation takes place without straining, the non-zero strains that result from differentiation of the measured displacements are indicators of the random errors imposed by the evaluation technique. The maximum amplitudes of these strains were 5.93×10^{-4} and 3.74×10^{-4} in the radial and tangential directions respectively, whereas the standard deviations were 2.25×10^{-4} and 1.52×10^{-4}. From a comparison with the results of the tensile test, presented in figure 3, it can be concluded that the measured strain amplitudes in our experiment are well above the noise level.

4.2 Evaluation of stiffness properties

The inverse finite element modelling was considerably simplified by reducing the number of independent parameters through eqs. (6)-(8). However, when solving the overdetermined system of equations provided by eq. (8), the obtained radial variations for D_{21} and D_{33} were not smooth, but rapidly oscillating from one point to another. This behaviour was particularly strong in the latewood region, which is probably a result of both the strain amplitudes and the tangential variation of the strains being small in this region. The oscillations were reduced by introducing two sets of additional penalty equations into the system of equations,

$$\left[D_{21}(r_{i-1}) - 2D_{21}(r_i) + D_{21}(r_{i+1}) \right] w_{21} = 0 \text{ and} \tag{10}$$

$$\left[D_{33}(r_{i-1}) - 2D_{33}(r_i) + D_{33}(r_{i+1}) \right] w_{33} = 0 \,, \tag{11}$$

where w_{21} and w_{33} are weights. By assigning non-zero values to these weights, any deviation from a piecewise linear variation of D_{21} and D_{33} will be penalised in the least-square solution of the overdetermined system of equations. Although this method helped in reducing the oscillations, no reliable results could be obtained for the stiffness properties at the two outermost sampling points in the latewood. For this reason, no data are presented for r>0.94 in figure 4.

The evaluation of stiffness properties is thus not as straightforward as the strain field determination, and the results are certainly affected by the choice of cost-function in eq. (9) and the applied numerical technique in general. It is also difficult to quantify to what extent the results are influenced by a certain numerical procedure. Benchmark problems, where strain fields calculated from known stiffness properties are analysed, can be used for this purpose.

The tangential strain field, calculated by use of the determined stiffness variation in the growth ring, is shown together with the measured counterpart in figure 5. The large-scale variation across the growth ring is reproduced by the finite element model, and the deviations between measurement and prediction are largely due to local stiffness variations in the material.

Finally, it should be noticed that the coefficients of elasticity were determined on the assumption of plane stress ($\sigma_L = \tau_{RL} = \tau_{TL} = 0$). This assumption certainly conflicts with the real state of stress close to the sharp latewood-earlywood interface, where large out-of-plane stresses are expected, due to the abrupt change in material stiffness. The presented stiffnesses should therefore be considered as plane stress equivalent properties, and one must be careful when applying them in three-dimensional or plane strain models of the growth ring.

5. CONCLUSIONS

Digital speckle photography in combination with optical microscopy was successfully applied for determining the deformation field in a single growth ring of Norway spruce. The size of the investigated area in the radial-tangential plane was 2.44x5.40 mm, and the spatial resolution in the measurements was 67.2 μm in the radial and 134.4 μm in the tangential direction.

The in-plane strains were obtained by differentiation of the measured deformation field. From the strain fields, the in-plane coefficients of elasticity were determined with respect to radial position within the growth ring. A plane stress finite element model of the growth ring was used for iterative solution of the inverse problem, and the number of searched stiffness parameters could be significantly reduced by directly applying the equations of equilibrium to the measured strain fields. Somewhat surprisingly, it was possible to determine E_R, E_T, G_{TR} and ν_{TR} from a single uniaxial tension test, in which the load was applied in the radial direction. The explanation is

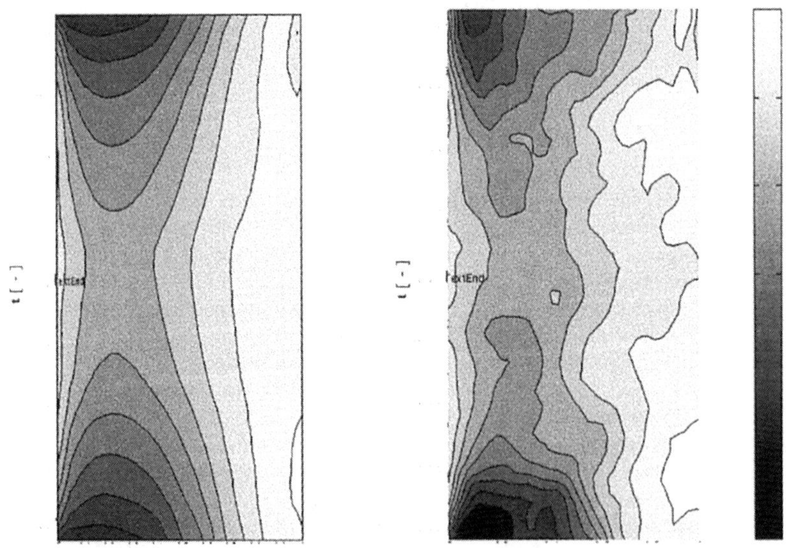

Fig 5. Calculated and measured distribution of tangential strain in the growth ring. The depicted area is the same as in figure 3, and the unit is millistrain (10^{-3}).

found in the abrupt change in stiffness from latewood to earlywood, which in itself introduced a biaxial variation in stress and strain in the test specimen. However, the determination of stiffness properties from measured strain fields was not trivial, and the results depend in part on the applied evaluation technique.

Acknowledgements

Financial support from the Swedish Foundation for Strategic Research (SSF) for L.O. Jernkvist is gratefully acknowledged. We are also indepted to Dr. M. Sjödahl and Prof. L. Berglund at Luleå University of Technology, who kindly placed the DSP algorithm and experimental equipment to our disposal.

References

1. Aghdam, M.M., Smith, D.J. and Pavier, M.J. Finite element micromechanical modelling of yield and collapse behaviour of metal matrix composites, *J. Mech. Phys. Solids*. **48** (2000) 499-528.
2. Gåsvik, K.J. *Optical Metrology*, John Wiley & Sons, 1995, Chichester, UK.
3. Synnergren, P. *White Light and X-ray Digital Speckle Photography*, Doctoral thesis, LTU-DT--00/12--SE, Luleå University of Technology, 2000, Luleå, Sweden.
4. Dinwoodie, J.M. *Wood - Nature's Cellular, Polymeric Fibre-Composite*, The Institute of Metals, 1989, London, UK.
5. Thuvander, F., Jernkvist, L.O. and Gunnars, J. Influence of repetitive stiffness variation on crack growth behaviour in wood, *J. Mat. Science*. **35**(24) (2000) 6259-6266.
6 Kahle, E. and Woodhouse, J. The influence of cell geometry on the elasticity of softwood, *J. Mat. Sciences*. **29** (1994) 1250-1259.
7 Sjödahl, M. Electronic speckle photography: increased accuracy by nonintegral pixel shifting, *Appl. Opt.* **33**(28) (1994) 6667-6673.
8 Sjödahl, M. Accuracy in electronic speckle photography *Appl. Opt.* **36** (1997) 2875-2885.
9 Jernkvist, L.O. and Thuvander, F. Experimental determination of stiffness variation across growth rings in *Picea abies*, *Holzforschung*. **55**(3) (2001) 309-317.

IDENTIFICATION OF ELASTIC/VISCO-PLASTIC MATERIAL CONSTANTS UNDER IMPACT LOADING

Shigeru Aoki, Kenji Amaya, Tomohiro Noguchi
Department of Mechanical and Environmental Informatics Tokyo Institute of Technology, Tokyo 152-8552, Japan

Per Ståhle, Karl Gustav Sundin
Division of Solid Mechanics, Luleå University of Technology, SE-971 87 Luleå, Sweden

Abstract A numerical simulation is performed to identify the material parameters of an elastic/visco-plastic material under tensile impact loading. The identification is based on an inverse analysis using the Kalman filter technique and the finite element method. The input data to the inverse analysis are the time histories of the strains on specimens. The measurement errors can be treated reasonably by the Kalman filter technique, and the finite element method enables us to employ arbitrarily shaped specimens. Notched specimens are employed to obtain high strain rates at the notch root. It is found that an additional test under a relatively low strain rate is needed together with a test under a high strain rate to increase the accuracy of identification. The optimum strain rate in the additional test is determined by minimizing the largest eigen value of the covariance matrix of the estimated parameters.

1. INTRODUCTION

The effect of strain rate on the deformation of materials has been intensively investigated by many researchers. It is recognized that metal can be modelled as elastic/visco-plastic material, and various constitutive equations based on this model have been proposed [1–4]. For any constitutive equation, accurate determination of their parameters is important.

Usually they must be measured experimentally, but precise values are often difficult to obtain. For example, it is difficult to reduce the strain rate change during an impact (especially compressive) test or to eliminate the effect of the non-uniform stress distribution in a specimen induced inevitably due to the stress wave propagation. In many cases, complex and costly experimental

P. Ståhle and K.G. Sundin (eds.), IUTAM Symposium on Field Analyses for Determination of Material Parameters - Experimental and Numerical Aspects, 63–80.

procedures are required. There is a need for a simple and effective method to determine unknown material parameters.

In the previous paper [5], to overcome the above-mentioned difficulties, the problem to determine the elastic/visco-plastic material parameters was regarded as an inverse problem where the material parameters are identified from the output of the strain gauges on the input and output bars in the split Hopkinson's bar compressive test, and the problem was solved by using the Kalman filter technique [6–8] and the finite element method.

In this paper, a numerical simulation is performed to determine the elastic/visco-plastic material parameters under tensile impact loading by applying the method based on the inverse analysis. Because the finite element method enables us to employ arbitrarily shaped specimens, notched specimens are employed to obtain high strain rates at the notch root. Accuracy of identification is improved by adding test data with lower strain rate to minimize the largest eigen value of the covariance matrix of the estimated parameters [9].

2. SYMONDS AND TING'S CONSTITUTIVE EQUATION

The stress-strain relation of an elastic/visco-plastic material for uniaxial tension is generally given by

$$\dot{\varepsilon} = \frac{\dot{\sigma}}{E} \qquad : \sigma \leq \sigma^Y \tag{1}$$

$$\dot{\varepsilon} = \frac{\dot{\sigma}}{E} + g(\sigma, \varepsilon) \qquad : \sigma > \sigma^Y \tag{2}$$

where, σ and ε are the stress and the strain in uniaxial tension, respectively, and E and σ^Y are the Young's modulus and the yield stress for static loading, respectively. the dot $(\dot{\ })$ represents the time derivative.

The following function proposed by Symonds and Ting [10] is employed here as the visco-plastic strain rate $g(\sigma, \varepsilon)$

$$g(\sigma, \varepsilon) = 0 \qquad : \sigma \leq \sigma_f(\varepsilon) \tag{3}$$

$$g(\sigma, \varepsilon) = D\left(\frac{\sigma}{\sigma_f(\varepsilon)} - 1\right)^q \qquad : \sigma > \sigma_f(\varepsilon) \tag{4}$$

where D and q are material constants and $\sigma_f(\varepsilon)$ is the static flow stress. Equation (4) coincides with that by Malvern [11] for $q = 1$.

3. NUMERICAL SIMULATION OF PARAMETER IDENTIFICATION

We perform here a numerical simulation to estimate the parameters D and q in equation (4). Essentially, the procedure uses 'measured' strain record and the Kalman filter technique to infer the most suitable values. In carrying out this procedure, it is assumed that correct values for the other parameters, such as Young's modulus E, the Poisson's ratio ν, the tensile yield stress σ^Y and the static flow stress $\sigma_f(\varepsilon)$ are known a priori and that the only unknown constants in the constitutive model are D and q.

We consider the impact tensile test apparatus as shown in Figure 1 [12]. By employing a notched specimen as shown in Figure 2, a high strain rate about $10^4 - 10^5$[1/s] is obtained at the root of the notch. Figure 3 shows the time histories of the displacement at the left side of ligament $u_1(t)$ (See Figure 2, t : time) which were obtained from high speed photographs of steel (A533 grade B class 1) specimens subjected to impact loadings with two different hammer velocities.

First, we carry out a direct analysis in which the strain at the strain gauge location of the specimen is calculated from the $u_1(t)$ by assuming D and q to be given. Then, we add appropriate errors to the calculated strain to obtain the 'measured' strain. In the inverse analysis, using the 'measured' strains as input data, we estimate the values of the parameters D and q.

3.1. Direct analysis

In the direct analysis, we employ the finite element code ABAQUS. Figure 5 shows the finite element mesh of the half of the specimen where plane strain 4 node linear elements and infinite elements are used because the deformation is concentrated in the narrow and thick ligament(0.24mm × 3mm). We assume that the specimen is made of a steel(A533 grade B class 1) with the following material constants: $E = 200.1$ [GPa], $\sigma^Y = 300$ [MPa], $\nu = 0.3$ and mass density $\rho = 7.86 \times 10^3$ [kg/m^3]. The static flow stress $\sigma_f(\varepsilon)$ is shown as a dotted curve in Figure 4.

Because the deformation is concentrated in the ligament as stated above, the difference between the displacement at the left side of ligament $u_1(t)$ (See Figure 2) and that at the left edge of the above specimen (nodes between E and F in Figure 5) is negligibly small. Hence, the displacements obtained by interpolating the data $u_1(t)$ (Figure 3) are given to the nodes between E and F as the boundary conditions in the X-direction.

We arbitrarily set the visco-plasticity parameters to be $D = 2.0 \times 10^5$ [1/s] and $q = 5.0$ in the direct analysis. These are the values to be sought in the inverse analysis. The computed strains in the X-direction at the node C , ε_α and ε_γ, are shown as solid curves in Figure 6 and Figure 7, where the subscripts α and γ correspond to the two boundary conditions shown in Figure 3.

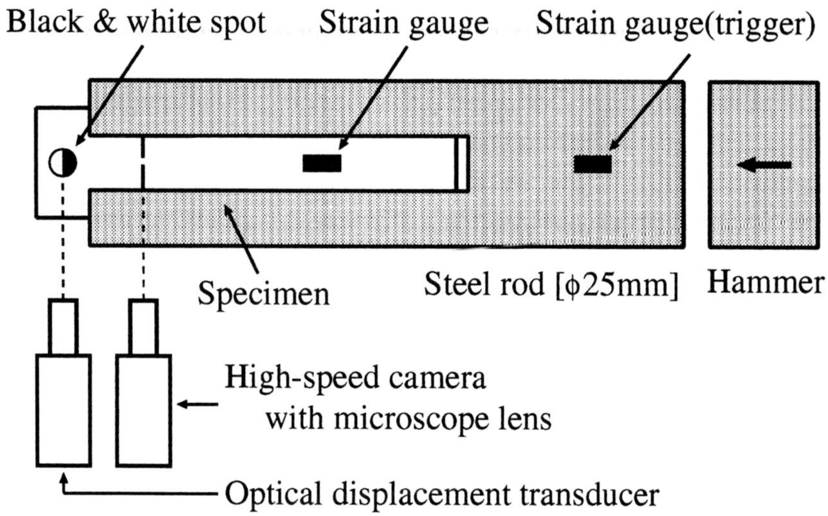

Figure 1 Experimental set-up

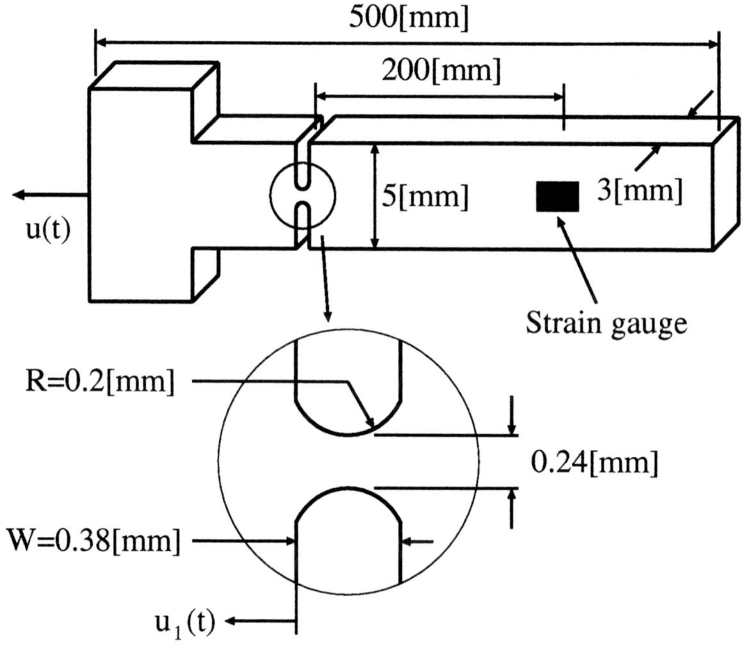

Figure 2 Specimen geometry

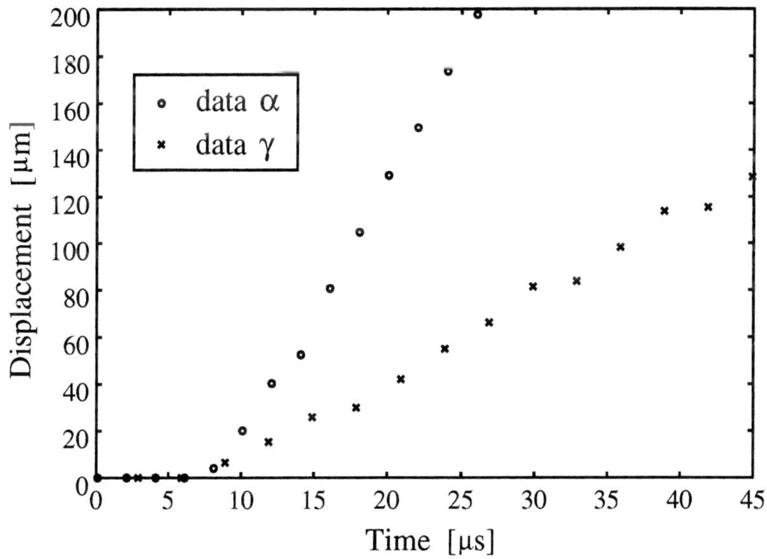

Figure 3 Measured time histories of displacement at the left side of the ligament $u_1(t)$ (average extension rates of the ligament are 11.1[m/s] and 3.7[m/s] for data α and γ, respectively)

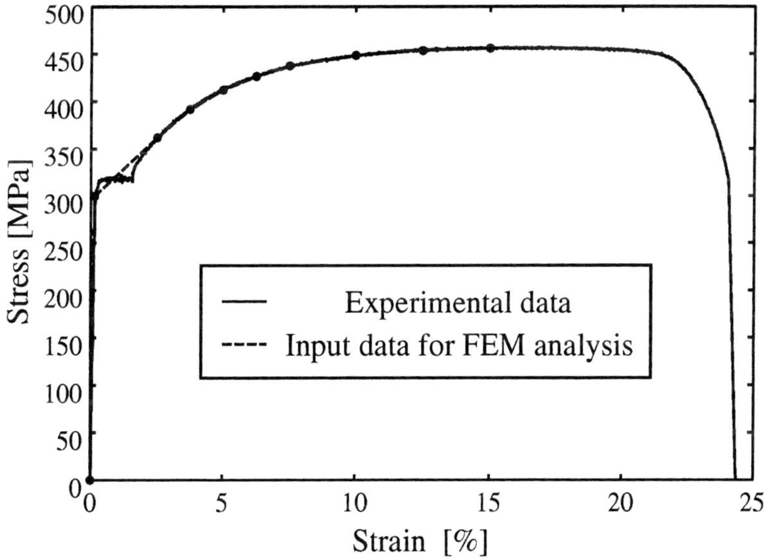

Figure 4 Stress-strain diagram (static)

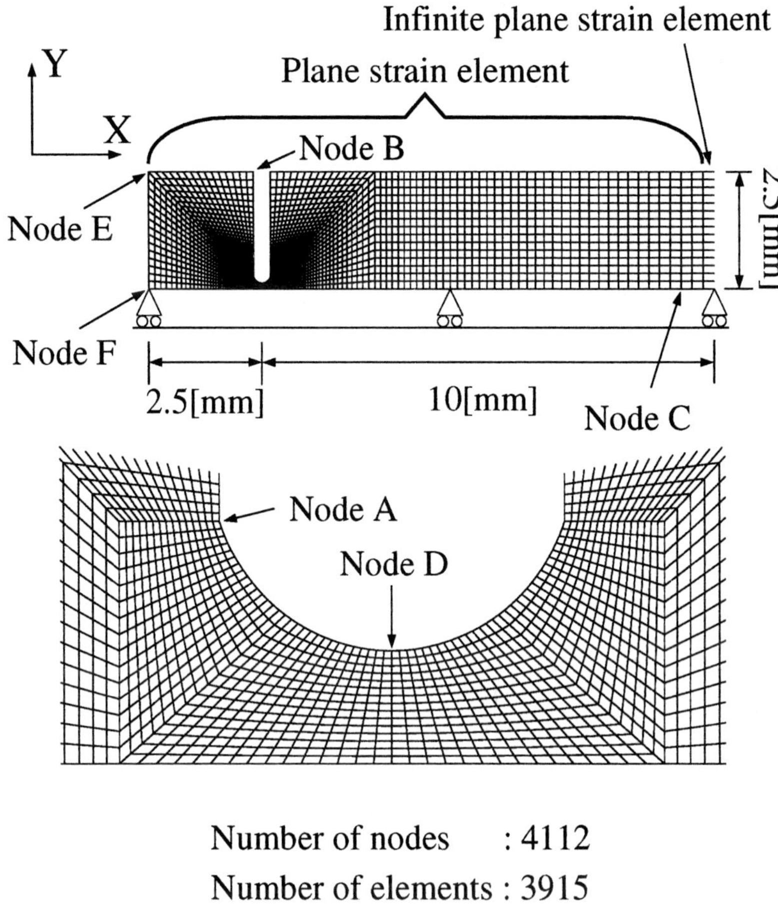

Figure 5 Finite element mesh

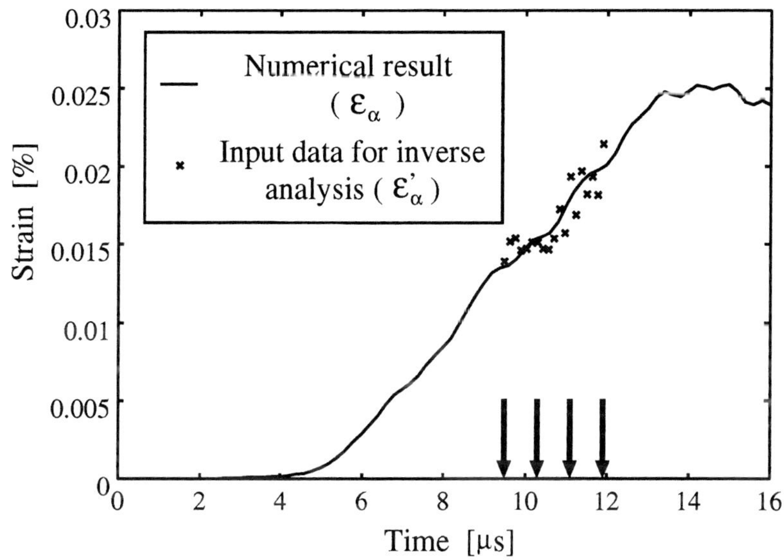

Figure 6 Time histories of strain at Node C (ε_α and ε'_α , $D = 2.0 \times 10^5$, $q = 5.0$)

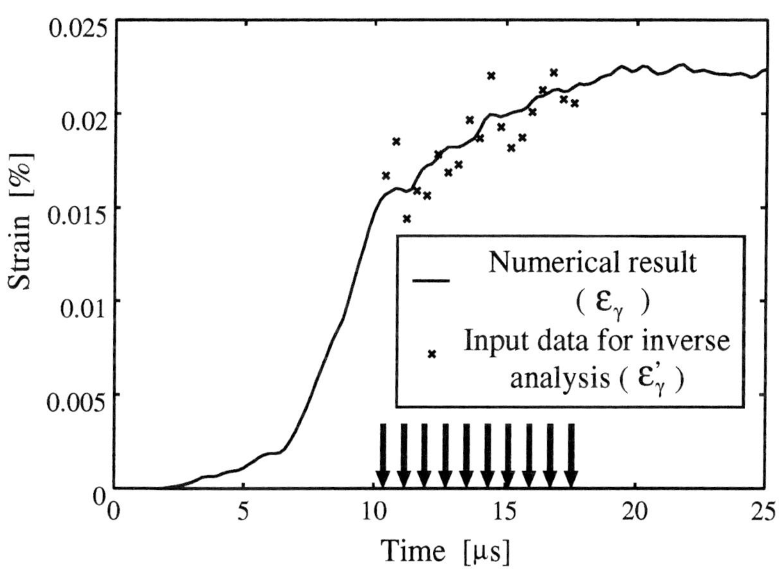

Figure 7 Time histories of strain at Node C (ε_γ and ε'_γ , $D = 2.0 \times 10^5$, $q = 5.0$)

3.2. Inverse analysis

In the inverse analysis, we employ the Kalman filter technique [13] (See Appendix), and assign $\boldsymbol{x}_s \equiv (D', q)^T$ to be the vector containing the unknown parameters as shown in equation (5), where $D' = \log_{10}(D/2)$. The subscript s indicates that the vectors can be (pseudo)-time or load increment dependent variables, and the superscript T is the transposition. We assume that the solution of the inverse analysis is $D = 2.0 \times 10^5$ [1/s] ($D' = 5.0$) and $q = 5.0$ which were used in the direct analysis.

$$\boldsymbol{x}_s = \begin{pmatrix} D' \\ q \end{pmatrix} = \begin{pmatrix} \log_{10}(D/2) \\ q \end{pmatrix} \quad (5)$$

The apriori information about D' and q is assumed to be considerably vague, and hence large diagonal values are assumed in the initial covariant matrix $\boldsymbol{P}_0$ as shown in equation (6). The range of possible initial estimate $\bar{\boldsymbol{x}}_0 (= \hat{\boldsymbol{x}}_0)$ is set as shown in equation (7).

$$\boldsymbol{P}_0 = \begin{pmatrix} 100 & 0 \\ 0 & 100 \end{pmatrix} \quad (6)$$

$$\bar{\boldsymbol{x}}_0 : \begin{cases} 3.0 \leq D' \leq 7.0 \\ 3.0 \leq q \leq 7.0 \end{cases} \quad (7)$$

We assign $\boldsymbol{z}_s$ to be the vector containing 'measured data', which are created by adding appropriate errors w to the direct analysis results ε_α and ε_γ (Figure 6 and Figure 7), as shown in equation (8). The errors w are generated as random values subject to a normal distribution(average=0, variance=$(1.0 \times 10^{-5})^2$), i.e., $p(w) \sim N(0, (1.0 \times 10^{-5})^2$.

$$\boldsymbol{z}_s = \begin{pmatrix} \varepsilon'_\alpha \\ \varepsilon'_\gamma \end{pmatrix} = \begin{pmatrix} \varepsilon_\alpha + w \\ \varepsilon_\gamma + w \end{pmatrix} \quad (p(w) \sim N(0, (1.0 \times 10^{-5})^2)) \quad (8)$$

The total number of the steps in the Kalman filter, L, is set to be 19 as shown in equation (9). The arrows indicate the measurement time, and the 'measured data', which are shown as crosses in these figures, are created by interpolating the direct analysis results at the measurement time and adding errors w.

The diagonal components of covariance matrix of measurement errors, $\boldsymbol{R}_s$, are assumed to be larger than w, because the input to the direct analysis, u_1 (Figure 3) must include another error. We assumed $(2 \times 10^{-5})^2$ for the 'data' at the measurement time, and $(4 \times 10^{-5})^2$ for the 'data' which were created through interpolation, as shown in equation (10).

$$L = 19 \quad (9)$$

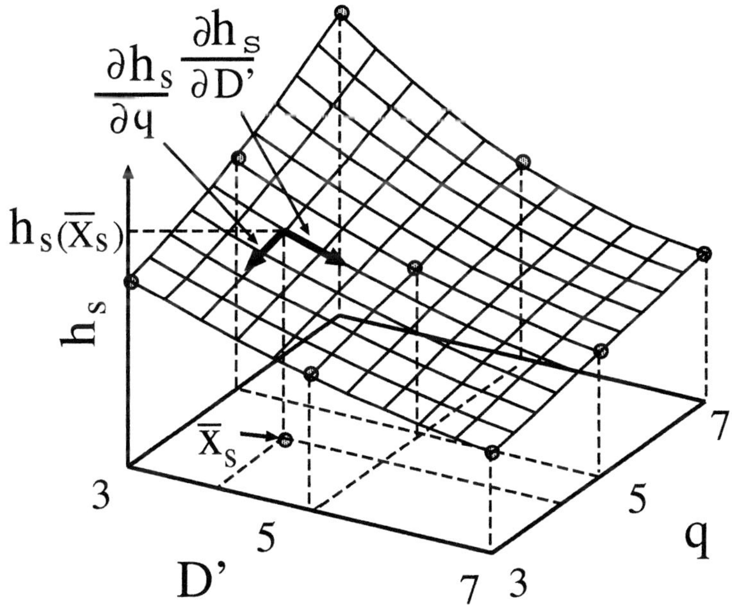

Figure 8 Lagrange's interpolation

$$\boldsymbol{R}_s = \begin{pmatrix} (2 \text{ or } 4)^2 & 0 \\ 0 & (2 \text{ or } 4)^2 \end{pmatrix} \times (10^{-5})^2 \qquad (10)$$

In the Kalman filter technique [13] (See Appendix), it is necessary to calculate $\boldsymbol{h}_s$ and its gradient $\boldsymbol{H}_s$ for various $\bar{\boldsymbol{x}}_s$'s. These functions must be determined by carrying out separate finite element calculations for various sets of D' and q. Since each calculation requires a significant computational time, we adopted the following procedure to minimize the computational effort, as in the previous paper [8]. Instead of carrying out the calculations for many different combinations of D' and q, we compute the strains ε_α and ε_γ for nine different sets of D' and q. Since the presumed range of D' and q forms a rectangular domain ($3.0 \leq D' \leq 7.0,\ 3.0 \leq q \leq 7.0$), we have chosen the nine sets at its end-points and mid-points (i.e, (D' , q)=(3.0, 3.0), (5.0, 3.0), etc.) , as shown in Figure 8. To estimate $\boldsymbol{h}_s$ at other points, as well as its gradient $\boldsymbol{H}_s$ within the D' and q domain, the bi-quadratic Lagrange's interpolation functions are employed.

We have carried out the Kalman filter with 11×11 different initial estimates $\bar{\boldsymbol{x}}_0$ (i.e.,D' and q are varied every 0.4). The estimated results of (D' , q) at the last step $s = 19$ are shown as crosses in Figure 9. It is observed that the

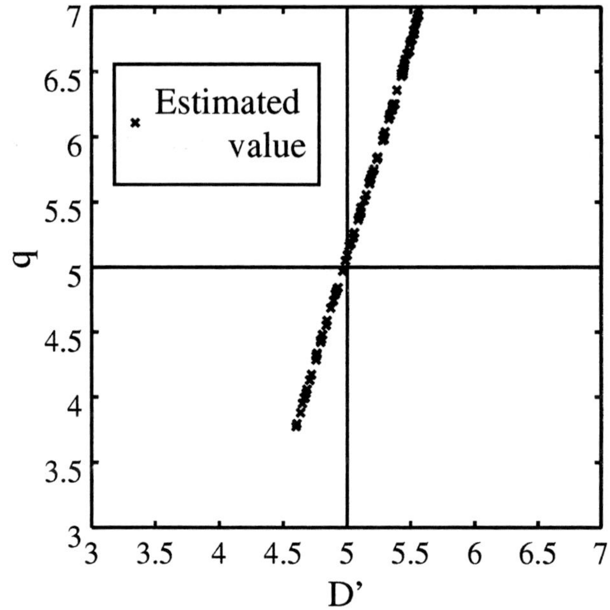

Figure 9 Estimated values of D' and q from various initial values (using ε'_α and ε'_γ)

results spread widely in a narrow region around the correct solution ($D' = q = 5.0$). Alternatively, they suggest that the input data to the inverse problem are not sufficient to estimate the correct parameters D' and q of the visco-plastic material.

4. IMPROVEMENT OF ESTIMATION ACCURACY

In order to improve the estimation accuracy, we considered to employ data with largely different strain rates. The problem here is to determine the optimum strain rates.

4.1. Optimum loading rate

We assume that equation (2) in Appendix hold generally, even if the subscript $(*)_s$ is deleted. Then, equations (3) – (7) hold without the subscript $(*)_s$, if we assume $\boldsymbol{x}$ and $\boldsymbol{z}$ to be general estimation and observation vectors, respectively. The component of $\boldsymbol{z}$ can be the strain with an arbitrary strain rate.

To improve the estimation accuracy under a limited number of measurement data (let m denote the number, i.e. the number of components of $\boldsymbol{z}$), it is

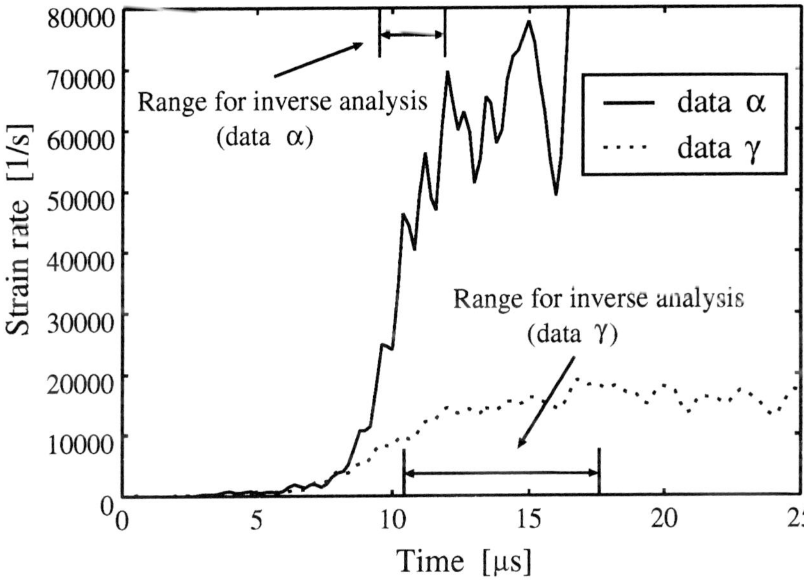

Figure 10 Time histories of strain rate at Node D (data α and data γ, $D = 2.0 \times 10^5$, $q = 5.0$)

desirable to minimize the maximum eigen value λ^2_{max} of the covariance matrix of estimation, $\boldsymbol{P}$ [9]. It is found from equation (7) that $\boldsymbol{P}$ depends on $\boldsymbol{H}$ and $\boldsymbol{R}$ if $\boldsymbol{M}$ is assumed to be constant. These matrixes $\boldsymbol{H}$ and $\boldsymbol{R}$ are fixed by determining the components of $\boldsymbol{z}$. It follows that λ^2_{max} can be minimized by selecting the m components of $\boldsymbol{z}$ appropriately.

Figure 10 shows the strain rates at the nodal point D (See Figure 5) for $D = 2.0 \times 10^5$, $q = 5.0$. It is seen that the strain rate in the time range used for the inverse analysis is $\dot{\varepsilon} = 2.0 \times 10^4 - 6.0 \times 10^4$[1/s] for data α, and $\dot{\varepsilon} = 1.0 \times 10^4 - 2.0 \times 10^4$[1/s] for data γ. Hence, setting $m = 2$ we employ $\dot{\varepsilon} = 4.0 \times 10^4$[1/s] or 1.0×10^4[1/s] as one of the two strain rates of $\boldsymbol{z}$, and seek the other strain rate which minimizes λ^2_{max}.

Although it is possible to determine the optimum strain rate exactly by calculating $\boldsymbol{H}$ with the finite element method, it would take a lot of computational time. We, therefore, employ an approximate but efficient method here. Neglecting the first term in the right hand side of equation (2), i.e., assuming that the elastic strain rate is negligibly small, we have from equations (2) and (4)

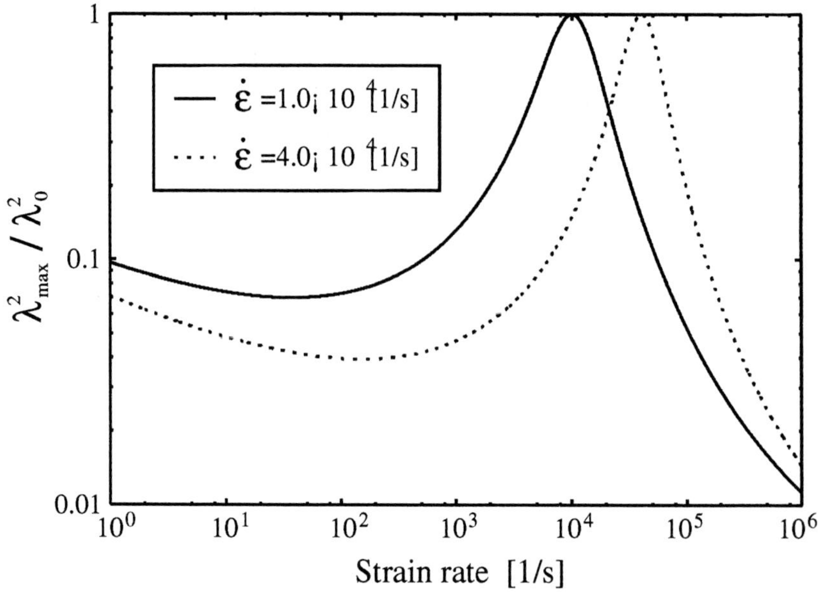

Figure 11 Effect of strain rate on the maximum eigen value of covariance matrix of estimated parameters ($\boldsymbol{P}$)

$$\frac{\sigma}{\sigma_f(\varepsilon)} = 1 + \left(\frac{\dot{\varepsilon}}{2 \times 10^{D'}}\right)^{1/q} \tag{11}$$

The covariance matrix $\boldsymbol{P}$ is given by deleting the subscript $(*)_s$ from equation (7) in Appendix.

$$\boldsymbol{P} = (\boldsymbol{M}^{-1} + \boldsymbol{H}^T\boldsymbol{R}^{-1}\boldsymbol{H})^{-1} \tag{12}$$

We take $\sigma/\sigma_f(\varepsilon)$ as a component of the observation vector, and assume that $\boldsymbol{M}$ and $\boldsymbol{R}$ in equation (12) are given by $\boldsymbol{P}_0$ (equation (6)) and $(0.1)^2\boldsymbol{I}$ ($\boldsymbol{I}$ is 2×2 unit matrix), respectively. $\boldsymbol{H}$ can be obtained by differentiating the right hand side of equation (11) with respect to D' or q for any fixed value of $\dot{\varepsilon}$. We fixed one strain rate of the two strain rates as $\dot{\varepsilon} = 1.0 \times 10^4$[1/s] or 4.0×10^4[1/s] as stated above, and obtained $\boldsymbol{P}$ and its eigen values as a function of the other strain rate. Figure 11 shows the normalized maximum eigen values λ^2_{max} / λ^2_0 versus the other strain rate, where λ^2_0 is the eigen value of $\boldsymbol{P}_0$.

It is found that a very large strain rate ($\dot{\varepsilon} > 10^6$[1/s]) is necessary to minimize λ^2_{max}. Since it is difficult to perform such a high loading rate test, we take

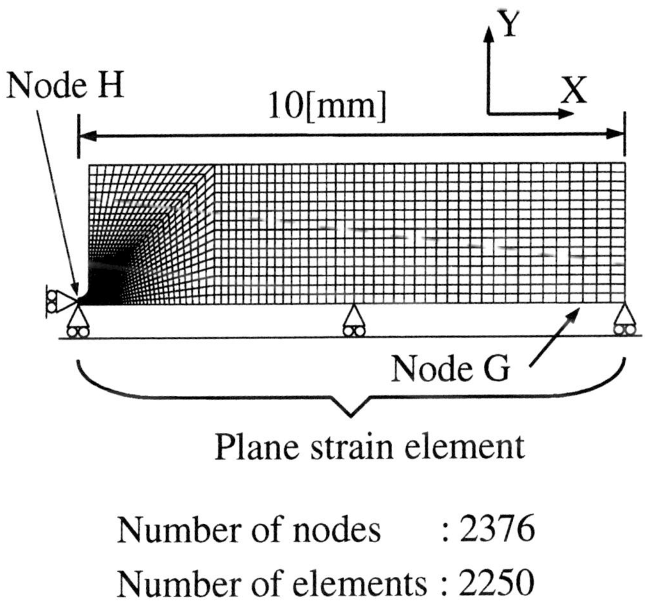

Figure 12 Finite element mesh for quasi-static test specimen

$\dot{\varepsilon} = 100$ [1/s], around which λ^2_{max} takes a local minimum, as the reasonable strain rate.

4.2. Estimation using new specimen

To obtain $\dot{\varepsilon} = 100$ [1/s], we consider a quasi-static tensile test using a new notched specimen, the main part of which has the same shape and size as those of the impact test specimen (Figure 2). We performed a direct analysis at first. Figure 12 shows the finite element mesh of a quarter of the new specimen taking symmetry into consideration, where plane strain 4-node linear elements are used.

Because there was only a small difference between the dynamic analysis result and the quasi-static one (i.e., the result without taking the inertia effect into account), we performed a quasi-static analysis. Figure 13 shows the time history of the strain rate at the node H (Figure 12) for the case where the right end of the specimen (Figure 12) with $D' = q = 5.0$ is pulled with the velocity 8.0 [mm/s]. It is observed that $\dot{\varepsilon}$ is about 100[1/s].

The time history of the strain in the X-direction at the node G (Figure 12), which is denoted with ε_δ, is shown in Figure 14. We created one component of the observation vector $\boldsymbol{z}_s$ by adding the error w shown in equation (8) to ε_δ, as shown in Figure 14. We took $\varepsilon'_\gamma (= \varepsilon_\gamma + w)$ as the other component of $\boldsymbol{z}_s$ as shown in equation (13).

$$\boldsymbol{z}_s = \begin{pmatrix} \varepsilon'_\gamma \\ \varepsilon'_\delta \end{pmatrix} = \begin{pmatrix} \varepsilon_\gamma + w \\ \varepsilon_\delta + w \end{pmatrix} \tag{13}$$

We carried out Kalman filter using the above observation vector. Figure 15 shows the estimated result of (D', q). It is found that a better estimation is obtained compared with the result in the previous chapter. Figure 16 shows the maximum and minimum eigen values of covariance $\boldsymbol{P}_{19}$ for the new (crosses) and old (circles) observation vectors. It is found that both of the maximum and minimum eigen values are lowered by employing the new observation vector shown in equation (13).

5. CONCLUSIONS

Since it is difficult to determine the parameters of the elastic/visco-plastic materials experimentally, it is desired to develop an efficient method. In this paper, a numerical simulation to identify the material parameters by applying an inverse analysis was performed. A tensile impact test was considered and the input data to the inverse analysis were the time histories of the strains on specimens. The Kalman filter and the finite element method were used in the inverse analysis. Because an arbitrarily shaped specimen is available in this method, a notched specimen was used to obtain a high strain rate at the notch root.

It was found from the simulation that accurate identification is difficult if only high strain rate data are used, i.e., other data with lower strain rate are necessary together with the high strain rate data. It was also found that the optimum strain rate to improve the identification accuracy can be determined by minimizing the largest eigen value of the covariance matrix of the estimated parameters.

Appendix: Kalman filter

The state transition and observation equations for the present problem are given by

$$\boldsymbol{x}_{s+1} = \boldsymbol{x}_s \tag{1}$$
$$\boldsymbol{z}_s = \boldsymbol{h}_s(\boldsymbol{x}_s) + \boldsymbol{v}_s \tag{2}$$

In the above, equation (1) states that the state vector $\boldsymbol{x}_s$ to be estimated is independent of step $s (s = 1, 2, \cdots, L)$. Also, equation (2) states that the observation vector $\boldsymbol{z}_s$ is expressed as a nonlinear function $\boldsymbol{h}_s(\boldsymbol{x}_s)$ plus observation

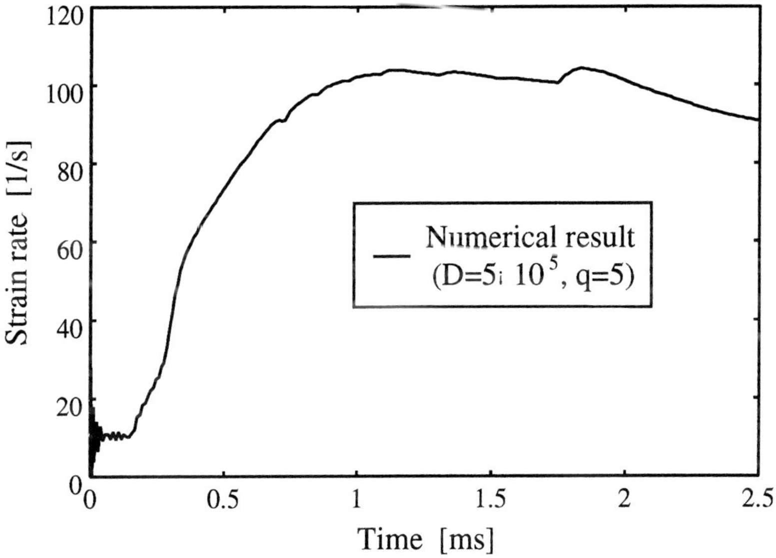

Figure 13 Time history of strain rate at Node H ($D = 2.0 \times 10^5$, $q = 5.0$)

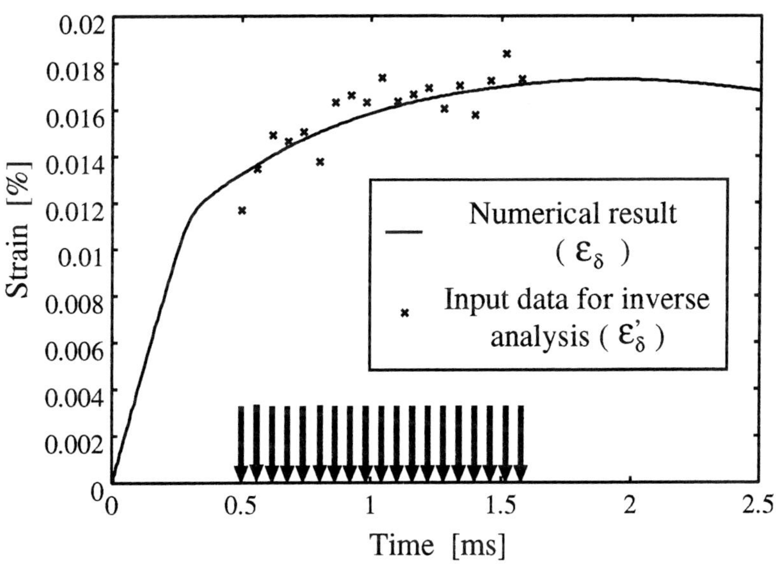

Figure 14 Time histories of strain at Node G (ε_δ and ε'_δ , $D = 2.0 \times 10^5$, $q = 5.0$)

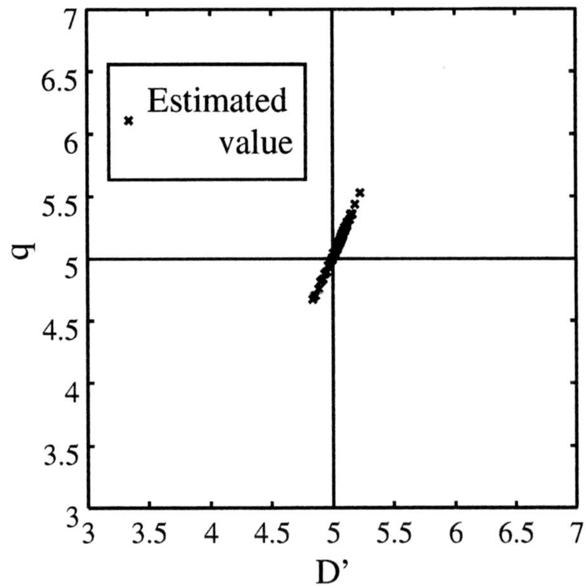

Figure 15 Estimated values of D' and q from various initial values (using ε'_γ and ε'_δ)

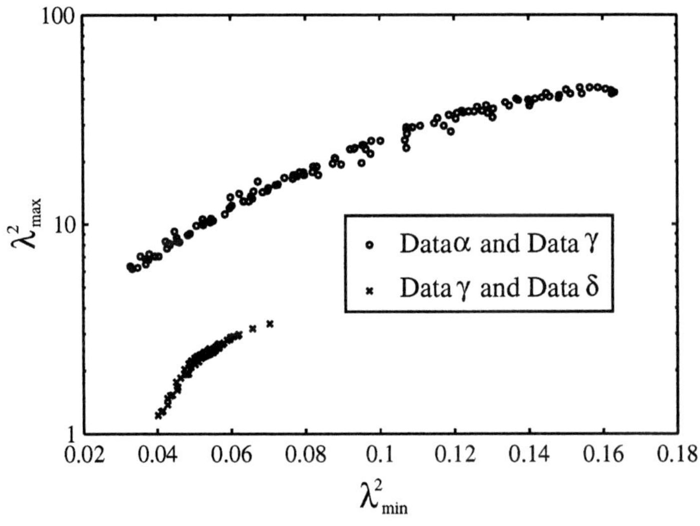

Figure 16 Improvement of eigen value of $\boldsymbol{P}_{19}$

error $\boldsymbol{v}_s$. The $\boldsymbol{v}_s$ and $\boldsymbol{x}_s$ are random variable vectors, which are independent of each other and subject to normal distributions. $\boldsymbol{R}_s$ (= the covariance matrix of $\boldsymbol{v}_s$), $\bar{\boldsymbol{x}}_0$ and $\boldsymbol{P}_0$ (= initial estimates of mean value and covariance of $\boldsymbol{x}_s$, respectively) are assumed to be given.

Let $\bar{\boldsymbol{x}}_s$ and $\boldsymbol{M}_s$ denote respectively the mean value and covariance matrix of $\boldsymbol{x}_s$ before taking account of $\boldsymbol{z}_s$, and let $\hat{\boldsymbol{x}}_s$ and $\boldsymbol{P}_s$ denote respectively the estimated value and covariance matrix of $\boldsymbol{x}_s$ after taking account of $\boldsymbol{z}_s$. If linear approximation of equation (2) around $\bar{\boldsymbol{x}}_s$ is made, then the $p(\boldsymbol{x}_s|\boldsymbol{z}_s)$, the probability density function after taking account of $\boldsymbol{z}_s$, is given as [13]

$$p(\boldsymbol{x}_s|\boldsymbol{z}_s) = \frac{1}{(2\pi)^{n/2}|\boldsymbol{P}_s|^{1/2}} \exp\left\{-\frac{1}{2}(\boldsymbol{x}_s - \hat{\boldsymbol{x}}_s)^T \boldsymbol{P}_s^{-1}(\boldsymbol{x}_s - \hat{\boldsymbol{x}}_s)\right\} \tag{3}$$

where n is the total number of the components of $\boldsymbol{x}_s$ and

$$\hat{\boldsymbol{x}}_s = \bar{\boldsymbol{x}}_s + \boldsymbol{K}_s(\boldsymbol{z}_s - \boldsymbol{h}_s(\bar{\boldsymbol{x}}_s) - \bar{\boldsymbol{v}}_s) \tag{4}$$

$$\boldsymbol{K}_s = \boldsymbol{P}_s\boldsymbol{H}_s^T\boldsymbol{R}_s^{-1} \tag{5}$$

$$\boldsymbol{H}_s = \left.\frac{\partial \boldsymbol{h}_s(\boldsymbol{x}_s)}{\partial \boldsymbol{x}_s}\right|_{\boldsymbol{x}_s=\bar{\boldsymbol{x}}_s} \tag{6}$$

$$\boldsymbol{P}_s = (\boldsymbol{M}_s^{-1} + \boldsymbol{H}_s^T\boldsymbol{R}_s^{-1}\boldsymbol{H}_s)^{-1} \tag{7}$$

Because the following equations hold for the state transition equation (1), and because equations (1) and (2) hold formally for $s = 0$, and because $\hat{\boldsymbol{x}}_0(= \bar{\boldsymbol{x}}_0)$ and $\boldsymbol{P}_0$ are given, the above procedures can be repeated .

$$\bar{\boldsymbol{x}}_{s+1} = \hat{\boldsymbol{x}}_s \tag{8}$$

$$\boldsymbol{M}_{s+1} = \boldsymbol{P}_s \tag{9}$$

References

[1] R.J.Clifton, Plastic Waves: Theory and Experiment, *Mechanics Today*, Pergamon, 102-167, (1972).

[2] K.Kishida, M.Nakano and Y.Sogabe, Experimental Study of Dynamic Plastic Constitutive Relation on Pure Copper, *J. Soc. Mat. Sci. Japan*, **43**-4 95, 1559-1563 (1994).

[3] M.Nakano, K.Kishida, Y.Yamauchi and Y.Sogabe, Evaluation of Constitutive Equation on Pure Copper by Experiment and Numerical Analysis of Plastic Wave Propagation under Incremental Torsional Impacts, *J. Soc. Mat. Sci. Japan* , **45**-7, 780-784 (1996).

[4] S.Tanimura, Dynamic Behavior of Materials and the Constitutive Equations,
Trans. Japan Soc. Metals **29**-5, 337-344 (1990).

[5] S.Aoki, K.Amaya and F.Terui, A New Method For Identifying Elastic/Visco-Plastic Material Constants, *Proc. 4th World Congress on Computational Mechanics*, 82/1-82/15 (1998).

[6] A.Murakami and T.Hasegawa, Back Analysis by Kalman filter-finite elements and optimal location of observed points, *Proc. 6th Int. Conf. on Numerical Methods in Geomechanics*, Innsbruck, 2051-2058 (1988).

[7] M.Hoshiya and E.Saito, Structural Identification by Extended Kalman Filter, *J. Engineering Mechanics, ASCE*, **112**-12, 1757-1770 (1984).

[8] S.Aoki, K.Amaya, M.Sahashi and T.Nakamura, Identification of Gurson's Material Constants by Using Kalman Filter, *Computational Mechanics*, **19** -6,501-506 (1997).

[9] K.Amaya and S.Aoki, Optimization of Measurements for Inverse Problem, *Proc. 3rd Int. Conf. on Inverse Problem in Engineering*, Port Ludlow, WA, 1-6 (1999)

[10] P.S.Symonds and T.C.T.Ting, Longitudinal Impact on Viscoplastic Rods – Approximate Methods and Comparisons, *J. Appl. Mech.*, **31**,611-620 (1964).

[11] L.E.Malvern, Plastic Wave Propagation in a Bar of Material Exhibiting a Strain Rate Effect, *J. Appl. Math.*, **8**, 405-411 (1951).

[12] L.G.Melin, P.Ståhle and K.G.Sundin, High strain rate tensile testing using microscopic high speed photography, *Proc. 11th Int. Conf. on Experimental Mechanics*, (1998), 175-179.

[13] T.Arimoto, *Kalman Filter*, Sangyo-Tosho, Pub., Tokyo, 53-88 (1992).

PARAMETER IDENTIFICATION FOR INVERSE PROBLEMS IN METAL FORMING SIMULATIONS

J.P. Kleinermann*, J.P. Ponthot**, M. Hogge[+]
*FRIA Ph.D. Student, **Senior assistant (corresponding author), [+] Prof & Head
LTAS-Continuum Mechanics and Thermomechanics University of Liège, 1 Chemin des Chevreuils, B52/3 B4000 Liège-1, Belgium

Abstract To improve the quality of numerical simulations results, accurate material behavior models are required. Moreover, more and more complex constitutive laws are being proposed to describe peculiar material properties. *Parameter identification* is an inverse problem taking place in material model development. It consists in evaluating the material parameters which exist in the chosen model, leading to the most accurate model, minimizing the difference between experimental results and Finite Element Method (FEM) simulations. Hence, the parameter identification problem can be formulated as an optimization problem. We propose, in this paper, to solve this optimization problem with eight optimization methods in order to compare their efficiency and robustness. The eight implemented methods come either from literature, such as conjugate gradient method, BFGS, Levenberg-Marquardt, etc., or from original developments, such as a modified GCMMA method and an optimization method combination technique. At last, an optimization method dedicated to parameter identification problem will be proposed.

1. INTRODUCTION

Nowadays, the computer simulation of metal forming processes using the finite element method (FEM) has proven its efficiency and usefulness. It allows one to test and compare several candidate processes avoiding use of heavy "trial and error" method. Nevertheless, as in each simulation method, the FEM consists in modeling reality, and requires the introduction of some models approximating the real behavior. The required input data are mainly geometry, mesh, non linear material behavior laws, loading cases, friction laws, etc. The computation of process evolution and final results from these data is called the *direct problem*. Since efficient numerical methods have already been developed, these direct problems using FEM have reached some level of maturity. It then becomes possible to solve more complex problems, called *inverse problems*. The goal of inverse problems is to determine input data for one or more of the direct problems leading to a given result, known to be a result of this direct problem.

Parameter identification is an inverse problem where the desired result is the most accurate material parameters set, leading to the best fit between experimental and FEM

P. Ståhle and K.G. Sundin (eds.), IUTAM Symposium on Field Analyses for Determination of Material Parameters - Experimental and Numerical Aspects, 81–99.

results. Hence, the result of an experiment that must highlight material behavior constitutes an additional data of this inverse problem. Because of experimental and modeling errors, this inverse problem has no exact solution. That is why it is formed as an optimization problem consisting in minimizing a given norm of the gap between these two results by adjusting material parameters defining the chosen material model.

In this work, we propose to introduce a mathematical formulation of this in-verse problem. Then, some appropriate optimization methods will be described in order to solve the resulting implicit non-convex and non-linear optimization problem. A new combination technique will be presented which improves the robustness and efficiency of the basic methods. Eight optimization methods are then proposed (five combined methods and three basic methods used as references). These methods will then be compared in some numerical results including identification of large strain elastoplastic and elastoviscoplastic models with non-linear hardening, as well as identification of hyperelastic models. Finally, these optimization methods will be compared statistically in terms of robustness, accuracy and time-consumption.

2. PARAMETER IDENTIFICATION

The development of inelastic material models may be divided into four steps. First, an experimental test with correctly chosen measured data must highlight material behavior. Then, a mathematical model must be formulated, based on these experimental results and physical knowledge. The third step consists of identifying material parameters that appear in the chosen model (parameter identification problem). Finally, the resulting model must be tested and validated.

In many situations, simple tests, e.g. creep, relaxation or cyclic tests are conducted such that high stresses, high strain rates, etc. are avoided. These conditions prove to be sufficient in order to justify the assumption of uniform distribution of the resulting strains, stresses,... within the whole volume of the specimen. However, the assumption of uniformity cannot often be verified experimentally. Typical examples, induced by geometrical effects, are necking of the sample in a tension test, or specimen barreling due to friction in compression tests [10].

To account for this inhomogeneity within the sample, strategies have been recently developed, by performing parameter identification using finite element simulations [4],[5],[10],... The key idea is to simulate the performed experiment, trying to adapt material parameters so that FEM results agree with the same results as the experimental ones. This problem is mathematically called an inverse problem and can be seen as an optimization problem where the objective function is the gap between experimental and FEM simulation results. Optimization variables are material parameters that appear in the proposed model.

At first, a mathematical definition of this gap between experiment and simulation results is required. The experimental result consists of several discrete measurements. Hence, we based our definition on the distance between each of these experimental points and the corresponding one (same X-axis value) of the FEM result (see fig.1b). As proposed in literature, we chose an Euclidean norm to meld all these distance values into one single objective function value. Moreover, in order to obtain an average relative ordinate difference between a point of the experimental curve and its corresponding point on the FEM curve), we defined the function to be minimize as follows (this supposes that experimental measurements errors are proportional to the measured value $u_i^{\exp}$:

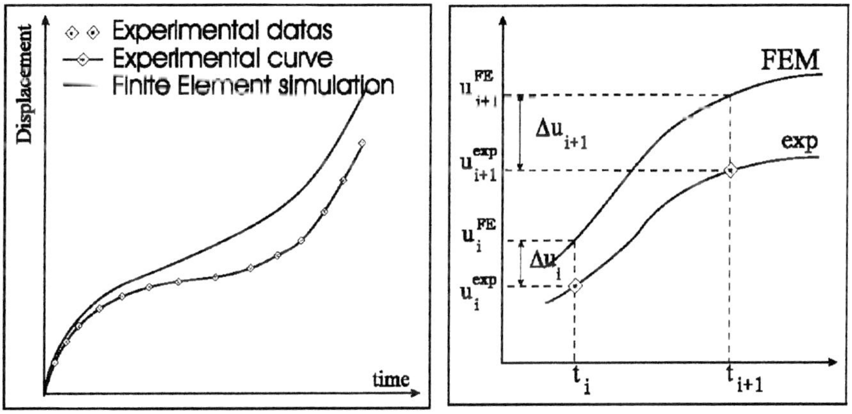

Figure 1 experiment-simulation gap (objective function) definition

$$S(\underline{x}) = \sqrt{\frac{1}{n}\sum_{i=1}^{n} P_i \left(\frac{u_i^{FEM}(\underline{x}) - u_i^{\exp}}{u_i^{\exp}}\right)^2} \qquad (1)$$

where $\underline{x}$ is the vector of parameter values, n is the number of experimental points, $u_i^{\exp}$ is the ordinate of the ith experimental point, u_i^{FEM} is the ordinate of the ith FEM point, and P_i is a weight factor attributed to the ith point.

3. OPTIMIZATION METHODS

This inverse problem needs an efficient optimization method able to minimize the implicit, non convex, and strongly non-linear objective function defined in (1). The above least-square objective function formulation is classical and admitted to be efficient. However, appropriate optimization methods used to solve this problem are difficult to find. Many different methods can be found in the literature to solve identification problems. These methods can be classified into three main categories: direct mathematical programming methods, linearization methods, and convex approximation methods.

3.1. Direct mathematical programming methods

Direct mathematical programming methods are methods classically used in structural optimization, such as *steepest descent, conjugate gradient, Newton method, or BFGS* [2] [12],... These can be used to minimize the implicit, non convex, and strongly non-linear objective function. At each iteration, these methods are divided into two main steps. At first, a descent direction ($\underline{s}^{(k)}$) is computed base on a gradient vector value ($\underline{g}(\underline{x}) = \partial S / \partial \underline{x}$) at the current iteration point ($\underline{x}^{(k)}$) and previous iterations results. These methods differ only in the formula for the descent direction computation.

$$\underline{s}^{(k)} = -\underline{g}^{(k)} \quad \text{stepest descent}$$

$$\underline{s}^{(k)} = -\underline{g}^{(k)} + \frac{\underline{g}^{(k)\mathrm{T}}(\underline{g}^{(k)} - \underline{g}^{(k-1)})}{\underline{s}^{(k-1)\mathrm{T}}(\underline{g}^{(k)} - \underline{g}^{(k-1)})}\underline{s}^{(k-1)} \quad \text{conjugate grad.} \tag{2a}$$

$$\underline{s}^{(k)} = -\left(\underline{\underline{R}}^{(k)}\right)^{-1}(\underline{g}^{(k)} - \underline{g}^{(k-1)}) \qquad \text{BFGS}$$

$$\text{where} \quad \underline{\underline{R}}^{(k)} = \underline{\underline{R}}^{(k-1)} + \frac{(\underline{g}^{(k)} - \underline{g}^{(k-1)})(\underline{g}^{(k)} - \underline{g}^{(k-1)})^{\mathrm{T}}}{\underline{s}^{(k-1)\mathrm{T}}(\underline{g}^{(k)} - \underline{g}^{(k-1)})} - \tag{2b}$$

$$- \frac{(\underline{\underline{R}}^{(k-1)}\underline{s}^{(k-1)})(\underline{\underline{R}}^{(k-1)}\underline{s}^{(k-1)})^{\mathrm{T}}}{\underline{s}^{(k-1)T}\underline{\underline{R}}^{(k-1)}\underline{s}^{(k-1)}}$$

Then, the objective function is minimized along this direction to find the new approximation point ($\underline{x}^{(k+1)}$) for the next iteration. This one-dimensional optimization problem is called "line-search" (cf. [2] [12]).

3.2. Gauss-newton and Levenberg-Marquardt methods

The Gauss-Newton (or Newton-Raphson) method is a very useful numerical computation method in order to seek roots of a non-linear equation or a non-linear equations system. Actually, in considered problems, optimum objective function value must be close to zero (equal to zero in ideal case). That is why our inverse problem can be formulated as finding the root of the non-linear equation system (for i=1,...,n):

$$u_i^{FEM}(\underline{x}) - u_i^{\exp} = 0 \tag{3}$$

Starting with the approximation parameters value $\underline{x}^{(k)}$, each iteration provides a new approximation $\underline{x}^{(k+1)}$, based on the first order Taylor approximation of the above system in $\underline{x}^{(k)}$. Then , $\underline{x}^{(k+1)}$ is computed in order to reach "$u_i^{FEM}(\underline{x}^{(k+1)}) - u_i^{\exp} = 0$". This linearization leads to an over determinated system:

$$\underline{\underline{J}}(\underline{x}^{(k+1)} - \underline{x}^{(k)}) = \underline{u}^{\exp} - \underline{u}^{FEM} \tag{4}$$

where
$$\underline{\underline{J}} = \frac{\partial \underline{u}^{FEM}}{\partial \underline{x}} \tag{5}$$

This equation is solved with a least-square method to give the new approximation point:

$$\underline{x}^{(k+1)} = \underline{x}^{(k)} - (\underline{\underline{J}}^{\mathrm{T}}\underline{\underline{J}})^{-1}\underline{\underline{J}}^{\mathrm{T}}(\underline{u}^{\exp} - \underline{u}^{FEM}) \tag{6}$$

This method has proven to be efficient when $\underline{x}^{(k)}$ is far from the optimum, but it cannot be used close to this optimum. Indeed, the true objective function value at the optimum is greater than zero, meaning that the above non-linear system (3) does not have any root. In this area the Gauss-Newton method becomes unstable and another optimization method must be used.

A way to improve that problem is to add a stabilization procedure. The Levenberg-Marquardt methods [5] [6] are similar to the Gauss-Newton method, introducing a New parameter, called Levenberg-Marquardt parameter λ, stabilizing the Gauss-Newton method. This parameter is updated after each iteration, to improve the convergence rate. The new iterative optimization formula becomes:

$$\underline{x}^{(k+1)} = \underline{x}^{(k)} - (\underline{\underline{J}}^{\mathrm{T}} \underline{\underline{J}} + \lambda \underline{\underline{I}})^{-1} \underline{\underline{J}}^{\mathrm{T}} (\underline{u}^{\exp} - \underline{u}^{FEM}) \qquad (7)$$

We can verify that with λ greater than zero, $\underline{x}^{(k+1)}$ is closer to $\underline{x}^{(k)}$ than with Gauss-Newton updating formula (6). Moreover, if one iteration lead to divergence, this iteration is recomputed by increasing the parameter . So, the new approximation $\underline{x}^{(k+1)}$ will be closer to $\underline{x}^{(k)}$ than the previous divergent iteration point. If this approximation still leads to a divergent k^{th} iteration, this iteration is recomputed again increasing the parameter λ. This sub-iterative process is used till convergence of the kth iteration.

Another way to stabilize the Gauss-Newton method is to combine it with a line-search [4]. The vector $\underline{x}^{(k+)}$-$\underline{x}^{(k)}$ chosen as a descent direction (see section 4.1) and a line-search is computed along this direction:

$$\underline{s}^{(k)} = -(\underline{\underline{J}}^{\mathrm{T}} \underline{\underline{J}})^{-1} \underline{\underline{J}}^{\mathrm{T}} (\underline{u}^{\exp} - \underline{u}^{FEM}) \qquad (8)$$

Line-search techniques are very complex and time consuming. Moreover, the direct programming methods described above seem to be more efficient than descent direction computation. That is why we considered only the basic Gauss-Newton method and its stabilization with Levenberg-Marquardt technique.

3.3. Convex approximations methods

Convex Approximation methods are recent optimization methods, much used in structural optimization problems. The key idea of these methods is to replace the basic problem with a sequence of simple explicit subproblems. The optimum of each subproblem becomes the new optimum parameter approximation, where a new explicit subproblem is formulated, approximating the basic problem. In general, both objective function and constraints must be approximated to form each subproblem, but in our inverse problems we do not consider only boundary and linear constrains, which do not have to be approximated because of their simple explicit form. That is why each explicit subproblem comes from the basic problem where the objective function $S(\underline{x})$ is replaced by its simple explicit approximation $S(\underline{x})$.

The first Convex Approximation method to be used was the reciprocal variables method, which replaces basic parameters $\underline{x}$ with reciprocal variables $z_i=1/x_i$ (for i = 1, ..n) in the optimization problem. Then, at each iteration of the optimization process, the approximated simple explicit sub-problem is given by the linearization of the basic problem (meaning objective function and constrains linearization if they have a complex or implicit form), expressed in terms of these reciprocal variables. This method showed, in several cases, excellent results attributed to the fact that some optimization constraints for linear elastic problems are much less non-linear in terms of reciprocal variables (e.g. bar truss weight optimization).

3.3.1 CONvex LINearization method. The CONvex LINearization method (CONLIN) [1] is based on the use of these reciprocal variables, but also direct variables, according to the sign of the first derivative of the approximated function. Each iteration has four steps. First, we compute the derivative of the objective function and constraints (if complex or implicit form) with respect to the optimization variables. Secondly, according to the sign of these derivatives, an automatic choice between direct and reciprocal variables is made for each function to be approximated independently. Then, the convex approximation is given by the first order Taylor series of the function expressed in term of these chosen variables. Finally, the simple optimization sub-problem is solved and its optimum becomes the new approximation to the basic problem optimum. CONLIN method is well used in structural optimization problems. Moreover, it has recently been found that the CONLIN method is also very efficient in shape optimization problems.

3.3.2 Method of Moving Asymptotes (MMA) A decade ago, Svan-berg [18] [20] presented a new general non-linear optimization method, particularly efficient in structural optimization. The *Method of Moving Asymptotes* is a generalization of the CONLIN method, where the linearization intermediate variables can be modified during the iterative process, in order to adjust the convexity of each approximation. Instead of just using direct and reciprocal variables, this method uses the intermediate variables $1/U_I - x_I)$ and $1/(x_I - L_I = 1,..,n)$ respectively, where U_i and L_i are user-selected parameters that can be modified from one iteration to the next. They are called moving asymptotes. As in CONLIN, the objective function and constraints are approximated by the first order Taylor series of the function expressed in terms of these intermediate variables. In order to achieve stable and fast convergence, an efficient updating technique is required for these asymptotes, iteration after iteration. Svanberg [18] proposed an empirical but efficient formula. Later, we will also describe a more complex technique called the two points updating method.

We point out an important property of both CONLIN and MMA. The objective function is approximated by a monotonic increasing function if the function derivative is positive, or by a monotonic decreasing function if this derivative is negative. Hence, if constraints are not approximated (e.g. because of their simple explicit form), this involves a non-convergence of both methods if the minimum point is located strictly inside the feasible domain (no active solved using these methods [22].

3.3.3 Globally Convergent Method of Moving Asymptotes (GCMMA) In 1995, Svanberg proposed [19] in 1995 a very simple modification of MMA to make it globally convergent. This method, called *Globally Convergent Method of Moving Asymptotes* still creates convex approximations of problem functions (objective function and constraints if necessary), but they present a "U" shape with a minimum between two vertical asymptotes (the moving asymptotes defined in MMA) for each optimization variable. A new convex approximation parameter appears, called ρ. It is modified from one iteration to another as in moving asymptotes updating.

Consider the kth iteration of the optimization process, with starting parameters approximation, $\underline{x}^{(k)}$. Suppose that $\underline{U}^{(k)}$ and $\underline{L}^{(k)}$, the moving asymptotes at the kth iteration, and $\rho^{(k)}$, are known (see updating method described later). After computing the derivative of $S(x)$ with respect to each optimization variable, the subproblem objective function, $S^{(k)}(\underline{x})$ is given by:

$$\tilde{S}^{(k)}(\underline{x}) = \sum_{i=1}^{n}\left(\frac{p_i^{(k)}}{U_i^{(k)} - x_i} + \frac{q_i^{(k)}}{x_i - L_i^{(k)}}\right) + r^{(k)} \tag{9}$$

where

$$p_i^{(k)} = \left(U_i^{(k)} - x_i\right)^2\left(\left(\frac{\partial S(\underline{x}^{(k)})}{\partial x_i}\right)_+ + \frac{\rho^{(k)}}{2}\left(U_i^{(k)} - L_i^{(k)}\right)\right)$$

$$q_i^{(k)} = \left(x_i - L_i^{(k)}\right)^2\left(\left(\frac{\partial S(\underline{x}^{(k)})}{\partial x_i}\right)_- + \frac{\rho^{(k)}}{2}\left(U_i^{(k)} - L_i^{(k)}\right)\right) \tag{10}$$

$$r^{(k)} = S(\underline{x}^{(k)}) - \sum_{i=1}^{n}\left(\frac{p_i^{(k)}}{U_i^{(k)} - x_i^{(k)}} + \frac{q_i^{(k)}}{x_i^{(k)} - L_i^{(k)}}\right)$$

where the symbols $(\,)_+$ and $(\,)_-$ respectively mean.:

$$(y)_+ = \begin{cases} y & \text{if } y > 0 \\ 0 & \text{if } y < 0 \end{cases} \qquad (y)_- = \begin{cases} -y & \text{if } y < 0 \\ 0 & \text{if } y > 0 \end{cases}$$

Let us look at the graphic representation of such a convex approximation in a simple problem with one optimization variable. Figure 2 shows the objective function $S(\underline{x})$ (solid line) and its convex approximation, $\tilde{S}^{(k)}(\underline{x})$ at point A (dotted line). We can see the U shape of the approximation, characteristic of GCMMA.

As we can see in figure 2, moving asymptotes and updating from one iteration to another is an important part of this new method. Moving asymptotes updating is similar to MMA-updating (see two point updating method). Regarding ρ, Svanberg proposed the following updating law that has already proven its efficiency in structural optimization [19]. From equation 9, the approximation function $\tilde{S}^{(k)}(\underline{x})$ can be expressed as

$$\tilde{S}^{(k)}(\underline{x}) = h^{(k)}(\underline{x}) + q^{(k)}d^{(k)}(\underline{x}) \tag{11}$$

where $h^{(k)}(\underline{x})$ and $d^{(k)}(\underline{x})$ do not depend on $\rho^{(k)}$. Let us define

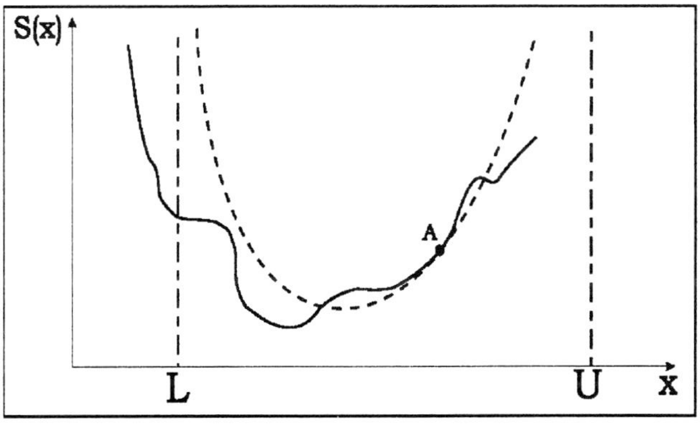

Figure 2 Convex approximation with GCMMA

$$\delta^{(k)} = \frac{S(\underline{x}^{(k+1)}) - \tilde{S}^{(k)}(\underline{x}^{(k+1)})}{d^{(k)}(\underline{x}^{(k+1)})} \tag{12}$$

The ρ updating rule proposed by Svanberg [19] is

$$\begin{aligned} \rho^{(k+1)} &= \rho^{(k)} \frac{6\rho^{(k)} + 4\delta^{(k)}}{6\rho^{(k)} + \delta^{(k)}} \quad \text{if} \quad \delta^{(k)} \geq 0 \\ \rho^{(k+1)} &= \frac{\rho^{(k)}}{2} \frac{2\rho^{(k)} - \delta^{(k)}}{\rho^{(k)} - \delta^{(k)}} \quad \text{if} \quad \delta^{(k)} < 0 \end{aligned} \tag{13}$$

3.3.4 Two-points updating method In order to ensure the quality of convex approximations with MMA and GCMMA, it is crucial to use an efficient moving asymptotes updating law. The key idea of the two-points updating [21] is to adjust moving asymptotes automatically to ensure that the kth iteration convex approximation will fit the exact function at the preceding iteration point $\underline{x}^{((k-1)}$, as shown in figure 3.

The moving asymptotes updating formula is given as:

$$\begin{aligned} L_i^{(k)} &= x_i^{(k)} - s\left(x_i^{(k-1)} - L_i^{(k-1)}\right) \\ U_i^{(k)} &= x_i^{(k)} + s\left(U_i^{(k-1)} - x_i^{(k-1)}\right) \end{aligned} \tag{14}$$

where the parameter s is computed to ensure that $\tilde{S}^{(k)}(\underline{x}^{(k-1)}) = S(\underline{x}^{(k-1)})$. The corresponding equation is formed by replacing equations (14) in the basic convex approximation equation (9), taking the implicit form

$$\tilde{S}^{(k)}(\underline{x}^{(k-1)}, s) = S(\underline{x}^{(k-1)}) \tag{15}$$

This non-linear equation is solved with a Newton-Raphson iterative scheme. When this parameter is computed, new moving asymptotes are evaluated from equations (14). The evolution of the convex approximation value at any admissible point $\underline{x}$ (i.e. placed between upper and lower moving asymptotes: $L_I < x_I < U_I$ for i=1,...,p) with respect to parameter s is presented on figure 4. We can verify by replacing equations (14) in (9)

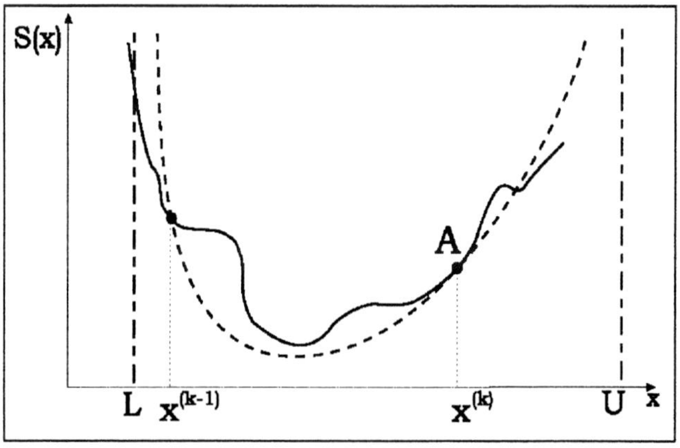

Figure 3 Convex approximation with GCMMA and two-point updating method

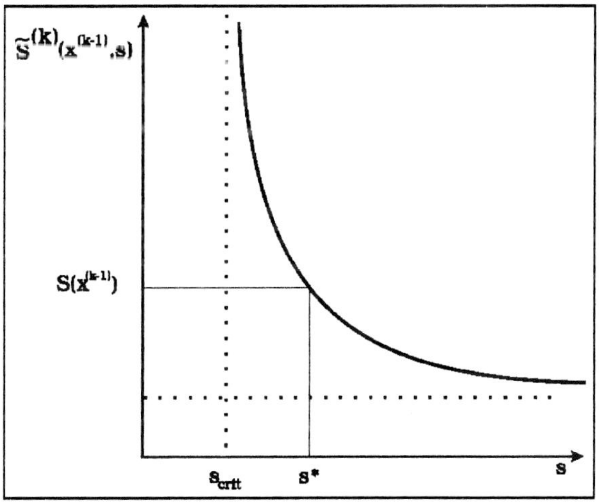

Figure 4 Convex approximation evolution with parameters

that:

$$\frac{\partial \tilde{S}^{(k)}(\underline{x},s)}{\partial s} \leq 0 \quad \text{and} \quad \frac{\partial^2 \tilde{S}^{(k)}(\underline{x},s)}{\partial s^2} \geq 0 \qquad (16)$$

3.3.5 Sub-iterative GCMMA method This GCMMA with two-point updating method was implemented in the large deformation finite element code METAFOR [14], and tested on inverse problems. This method led to fast convergence during an initial part of the iterations, but further iterations led to divergence. So, the resulting iterative optimization process could not converge.

That is why we propose modification of this method, avoiding these divergent iterations without disturbing the convergent ones. The key idea is to make sub-iterations when convex approximation parameters ($\underline{U}$, $\underline{L}$ and ρ) are unadapted, in order to modify them till they lead to a convergent iteration. If $\underline{x}^{(k)}$ is the starting point of the kth iteration, the basic GCMMA algorithm with two points updating gives a convex approximation of the objective function fitting $S(\underline{x}^{(k-1)})$ (two-point updating method). The minimum of this convex approximation becomes the new approximation point of the optimum parameter values, $\underline{x}^{(opt)1}$ (see fig 5a). After computing the objective function value in this point, we can check if the kth iteration is convergent ($S(\underline{x}^{(opt)1}) < S(\underline{x}^{(k)})$). If not, this means that the convex approximation of this iteration was too much wide-mouthed. Hence, to create a more conservative convex approximation, we reconsider the kth iteration but we prescribe the new convex approximation to pass through the resulting divergent point $\underline{x}^{(opt)1}$ (see fig 5b). So, the new minimum of this more conservative approximation, $\underline{x}^{(opt)2}$, will be closer to $\underline{x}^{(k)}$. If this new feasible point still leads to a divergent k^{th} iteration, another convex approximation is realized passing through this latest minimum point $\underline{x}^{(opt)2}$. This sub-iterative process is used till convergence of the kth iteration. This last convergent point will then be used as a starting point for the next iteration, $\underline{x}^{(k+1)}$.

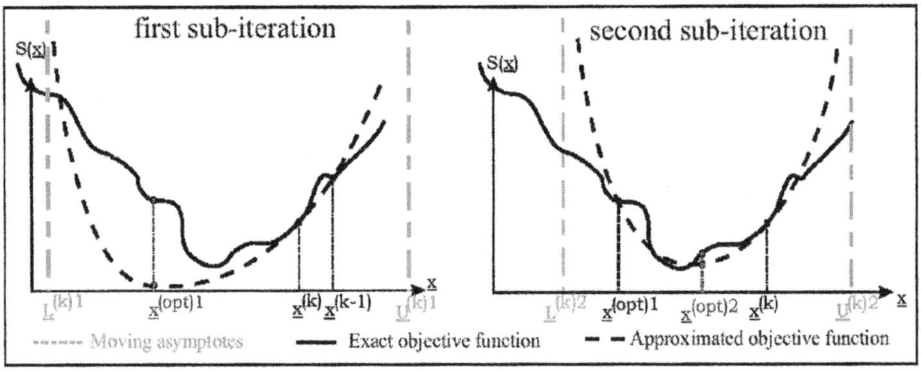

Figure 5 Sub-iterative method with GCMMA

3.4. Combined methods

The first computed numerical applications clearly showed the main difficulties in inverse problems. The objective function shape can be very different from one problem to another. That is why a very efficient optimizationmethod in one problem could be inefficient in another. Moreover, one method can improve some iterations of the optimization process but not in others. As described before, linearization methods (Gauss-Newton and Levenberg-Marquardt) can provide a quickly objective function decrease when the parameter set is far from the optimum value, but may be inefficient close to the optimum. Near the optimum, conjugate gradient method shows excellent results. That is why we naturally combined these methods to form *Gauss-Newton + conjugate gradient* and *Levenberg-Marquardt + conjugate gradient* methods. When the first method gives a rate of decrease of the objective function lower than a defined value, the iterative optimization process automatically shifts to the conjugate gradient method.

Continuing with this idea of optimization methods combinations, we tried to develop a robust optimization method able to solve any application, using the optimization method most likely to be efficient. Sub-iterative GCMMA showed, in many applications, a very low convergence rate when approaching the optimum, reaching this point very slowly. That is why we propose the combined method *GCMMA + conjugate gradient*. In the same way, in order to take advantages of Levenberg-Marquardt and Gauss-Newton methods which are likely to give very quick convergence in the first iterations, we propose to use *Gauss-Newton + GCMMA + conjugate gradient* and *Levenberg-Marquardt + GCMMA + conjugate gradient* methods.

In order to test and compare all these proposed methods on several numerical applications in terms of robustness, accuracy and CPU time, we implemented the eight following methods (3 reference methods + 5 combined methods). Each numerical application was solved with all these methods in order to compare their results.

1) Conjugate gradient (C)
2) BFGS (B)
3) GCMMA (G)
4) GCMMA + conjugate gradient (GC)
5) Gauss-Newton + conjugate gradient (NC)

6) Levenberg-Marquardt + conjugate gradient (LC)
7) Gauss-Newton + GCMMA + conjugate gradient (NGC)
8) Levenberg-Marquardt + GCMMA + conjugate gradient (LGC)

4. NUMERICAL APPLICATIONS

4.1. Identification of a non-linear hardening model

Consider the necking of a cylindrical beam. The geometry and mesh are shown in on figure 6. Note that the diameter in the middle of the beam is slightly smaller than at the end. This difference will initiate the necking. The external load is applied through a forced displacement of the beam ends, so that, at the end of the experiment, the total elongation is 14 mm.

The experimental result [13] is the external applied force evolution with beam elongation, and the necking radius evolution with elongation. The material is a special steel used in the piping manufacture for nuclear industry (Steel A-533 Grade B Class 1). To simulate its behavior, we chose an elastoplastic model with non-linear hardening [17]. This hardening is described by the law:

$$\tilde{S}^{(k)}(\underline{x}) = \underline{\sum_{i=1}^{n}\left(\frac{p_i^{(k)}}{U_i^{(k)} - x_i} + \frac{q_i^{(k)}}{x_i - L_i^{(k)}}\right)} + r^{(k)} \tag{17}$$

where $\bar{\varepsilon}^{pl}$ is the equivalent plastic deformation $\sqrt{\frac{2}{3}\varepsilon_{ij}^{pl}\varepsilon_{ij}^{pl}}$ and ε_{ij}^{pl} is the plastic deformation tensor.

This elastoplastic model with non-linear hardening contains 6 parameters. Elastic parameters are supposed to be known (*E = 206900N/*mm^2 and $\nu = 0.29$). The resulting identification problem is formed of a 4 parameters identification ($\sigma_{VM}^0, \sigma_{VM}^\infty$, δ and ζ), with an experimental result described by two evolution curves, respectively force and necking radius evolution with beam elongation.

The starting values and optimum parameter set, corresponding to the minimum gap between experiment and finite element simulation, are presented on the table. The corresponding simulation results, compared to the two experimental curves, are shown in figure 7. We see that initial parameters values give a

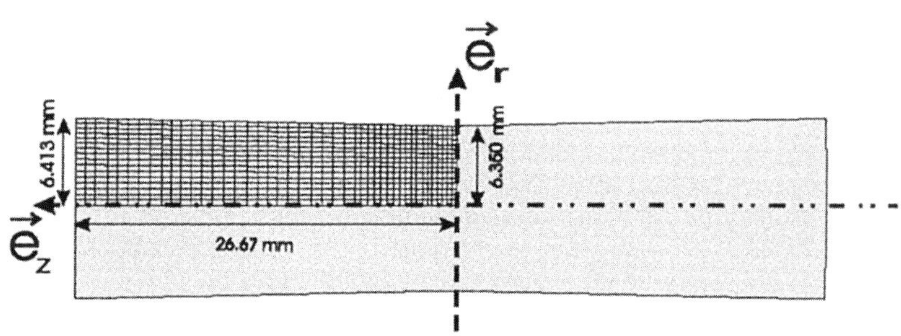

Figure 6 Initial geometry of the beam

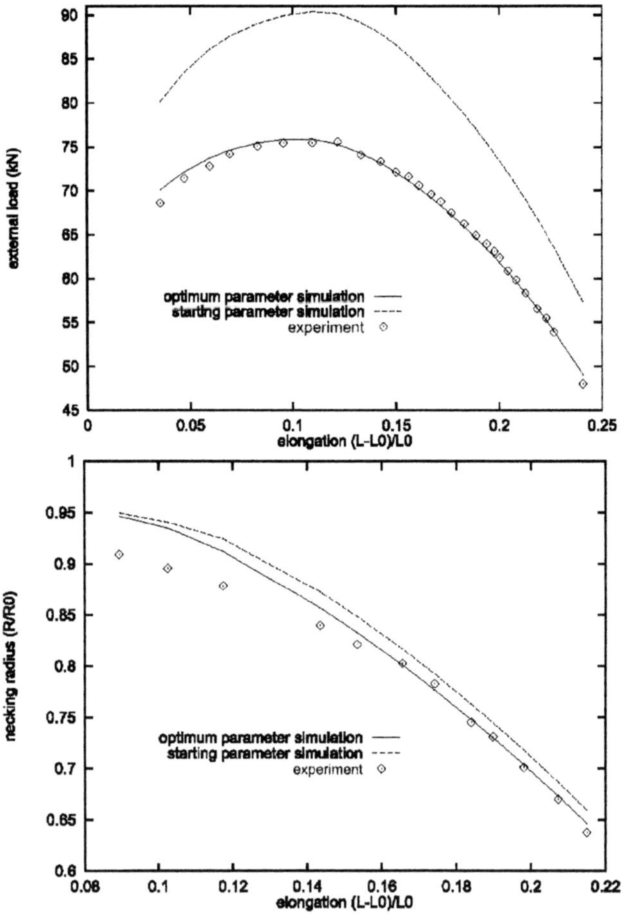

Figure 7 Experience - simulation results comparison

	σ_{VM}^{0} (N/mm²)	σ_{VM}^{∞} (N/mm²)	δ	ζ	S (%)
Initial values	500.0	800.0	25.000	300.0	15.697
Optimum	458.5	657.7	18.868	311.4	1.401

good approximation for necking radius evolution, but strongly overestimate the applied load. After parameter identification, the simulated applied load almost perfectly fits the experimental result. The necking radius evolution is well simulated for elongations greater than 0.15 but FEM simulation overestimates this radius for smaller elongations. This may be due to some other inaccurate model or experimental errors (for zero elongation, necking radius R/R_0 seems different from 1) because whatever parameter set is chosen, no simulation could fit radius evolution in this area.

The next table compares the results reached with the eight different optimization methods. We see that three different optima are reached, characterized by different δ

values, which is the least sensitive parameter. Modifying the parameters is a good way to distinguish optimization methods types. The Gauss-Newton method is not stabilized and tends to modify such parameters too much. Actually, Gauss-Newton method passes over the δ optimum value (about 18.9) to reach a local minimum withδ value about 16.7. The Levenberg-Marquardt method which can be seen as a stabilization of Gauss-Newton method seems to lead to the optimum δ value. Gradient based methods such as GCMMA, conjugate gradient and BFGS show a negligible decrease of δ. Gradient vector being almost parallel to the δ axis, this parameter is almost unmodified. That is why these methods lead to a local minimum corresponding to a δ value very close to the starting value (about 25). So, the best optimum seems to be given by the two methods using Levenberg-Marquardt. Comparing optimum values and minimum objective function value, we can say that LGC and LC seems to be the most accurate ones, but LGC is significantly faster.

4.2. Identification of a rubber hyperelastic model

In order to simulate the crushing of car door joint, a rubber hyperelastic model had to be identified using tension tests on the joint rubber [15]. Rubber is characterized by concepts of hyperelasticity and quasi incompressibility.

	σ_{VM}^{0} (N/mm^2)	σ_{VM}^{∞} (N/mm^2)	δ	ζ	S (%)	CPU (min)
C	391.4	651.8	24.999	329.1	1.411	54
B	401.0	649.5	24.999	337.2	1.421	114
G	415.3	661.9	24.996	329.6	1.833	56
GC	392.0	651.7	24.995	329.6	1.411	**47**
NC	480.6	660.0	16.691	306.4	1.431	117
LC	458.5	657.7	18.868	311.4	**1.401**	109
NGC	480.8	659.7	16.696	307.4	1.431	86
LGC	458.9	657.0	18.871	313.1	**1.402**	63

A hyperelastic material is defined by a strain energy rate ψ which is similar to the Helmholtz energy of the system. So, this energy function directly binds stresses and strains in the system. An important property of these materials is their ability to reach large deformations without irreversible phenomena.

When using isotropic material as considered here, we use a reduced strain rate tensor, where volumetric variation has been suppressed [16]. This form allows us to separate deviatoric and volumetric parts of the strain energy rate function as

$$\psi(\tilde{I}_1,\tilde{I}_2,J) = U(\tilde{I}_1,\tilde{I}_2) + W(J) \tag{18}$$

where U is the strain energy rate function of the incompressible material $W(J)$ acts like a penalty coefficient enforcing incompressibility $\tilde{I}_1$ and $\tilde{I}_2$ are the two first invariants of the so called reduced Finger tensor $\underline{\underline{\tilde{B}}}$.

Different hyperelastic models can then be expressed using different U function forms. We propose to identify two models and to compare their results. These are as follows:

- Gent-Thomas model

$$U(\tilde{I}_1,\tilde{I}_2) = C_1(\tilde{I}_1 - 3) + 3C_2 \ln \tilde{I}_2 \tag{19}$$

- improved Hart-Smith model

$$U(\tilde{I}_1,\tilde{I}_2) = C_1 \int \exp\left(C_3(\tilde{I}_1 - 3)^{C_4}\right) d\tilde{I}_1 + 3C_2 \ln \tilde{I}_2 \tag{20}$$

where C_1, C_2, C_3 et C_4 are material parameters to be identified. These models require identification of 2 and 4 parameters respectively.

4.2.1 Gent-Thomas model identification The next table presents optimum parameters values obtained with the eight proposed optimization methods. The starting point is $C_1 = 6\ N/mm^2$, $C_2 = 2$ N/mm^2. We see that, except for GCMMA and BFGS which are not efficient in this problem, each method converges to the optimum with different convergence rates. The last column presents the number of direct problem simulations (*nbdir*) that would lead to the same computation time as the inverse problem. It is calculated as computation time of the inverse problem divided by an average computation time for the corresponding direct simulation of the experiment.

	C_1 (N/mm^2)	C_2 (N/mm^2)	S (%)	CPU (min)	nbdir
C	391.4	651.8	3.282	1 h 54 m	180
B	401.0	649.5	3.455	11 h 41 m	1107
G	415.3	661.9	79.87	2 h 51 m	270
GC	392.0	651.7	3.335	2 h 22 m	224
NC	480.6	660.0	**3.280**	**1 h 20 m**	**126**
LC	458.5	657.7	3.379	2 h 14 m	212
NGC	480.8	659.7	3.284	2 h 53 m	273
LGC	458.9	657.0	3.285	2 h 50 m	268

Regarding objective function value, the NC method result was chosen as the sought optimum. Moreover, this method is the quickest one. The corresponding simulation result is compared to the experimental one in figure 8. After we tested a few other starting points, we feel confident obtained result represents a global optimum.

4.2.2 Improved Hart-Smith model identification To improve the previous simple model, we tried to identify the more complex Hart-Smith model. This model contains 4 material parameters. The next table presents the optimums reached with the proposed optimization methods. The starting point is $C_1 = 0.23$ N/mm^2, $C_2 = -0.2$ N/mm^2, $C_3 = 0.005$ and $C_4 = 1.5$.

These results show once more the large differences between these optimization methods, particularly for the less sensitive parameters, such as C_4. The Gauss-Newton method tends to change C_4 too much leading to a too low optimum value of about

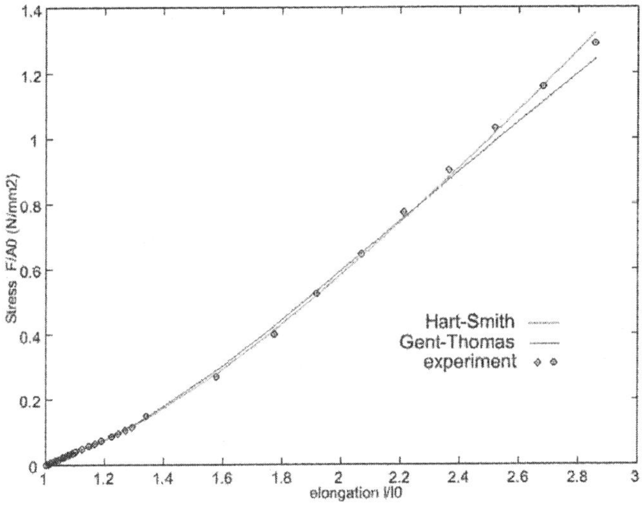

Figure 8 Experimental and simulations results comparison with the optimum models

0.3 N/mm^2. Levenberg-Marquardt method seems the most efficient method for updating C_4.

The other three methods (GCMMA, BFGS and conjugate gradient) hardly change C_4 at all, leading to an optimum value very close to the starting value. So, we can conclude that only LGC and LC methods reached the researched optimum. The corresponding simulation result is compared to the experimental one in figure 8. Others converge to local minima characterized by wrong C_4 values.

	C_1 (N/mm^2)	C_2 (N/mm^2)	C_3 (N/mm^2)	C_4 (N/mm^2)	S (%)	CPU (min)	nbdir
C	0.2766	-0.2244	0.00754	1.497	2.705	2 h 0 m	96
B	0.2428	-0.1902	0.01594	1.500	3.984	2 h 42 m	130
G	0.2380	-0.1856	0.02283	1.499	5.359	2 h 18 m	110
GC	0.2806	-0.2285	0.00664	1.495	2.689	**1 h 15 m**	**60**
NC	0.3010	-0.2494	0.00170	0.307	3.309	2 h 21 m	99
LC	0.2721	-0.2199	0.01409	1.166	**2.668**	1 h 46 m	85
NGC	0.2786	-0.2282	0.01987	0.339	3.837	1 h 24 m	67
LGC	0.2732	-0.2211	0.01396	1.166	**2.668**	1 h 58 m	95

LGC and LC can be divided into two main steps. The first iterations using Levenberg-Marquardt allow a good optimization of C_4 (because of its lower sensitivity) with, at the same time, an initial of the parameters C_1 to C_3. Then, after six iterations, the Levenberg-Marquardt method becomes less efficient. The second step of the process is performed using GCMMA and conjugate gradient methods. This step does not modify C_4 any more, but optimizes C_1 to C_3 precisely.

4.2.3 Comparison of these two optimum models Figure 8 compares the two models with the experimental results. We see that the improved Hart-Smith model fits the experimental result better than the Gent-Thomas model. Quantitatively, this corresponds to a objective function value, which measures gap between experimental points and simulation result, of 3.28% for Gent-Thomas, improved to 2.67 % with Hart-Smith.

Note that this difference appears mainly for elongation greater than 2.4. That is why, in order to test if using the more complex model (Hart-Smith) is useful, we simulated the car door crushing and compared stress results (see figure 9). Stress distribution along the joint is identical with both models, but stresses values are a little different. Maximum stress, for example, is 0.35 N/mm^2 with Gent-Thomas and 0.33 N/mm^2 with Hart-Smith. So, the Gent-Thomas model seems to overestimate stresses inside the door joint by about 7%. More tests and validation should be performed in order to compare these two models precisely, but it seems that using the improved Hart-Smith model is not worth the more complicated identification procedure and the doubled CPU time for simulation.

5. COMPARISON OF OPTIMIZATION METHODS

As we have already said, the main difficulty in these inverse problems is the objective function shape which can be very different from one problem to the next. That is why the most wanted property of an optimization method is robustness, meaning good convergence for most problems. As this property can only appears with statistics based on results for several inverse problems, figure10 presents optimization results of eleven numerical applications [7], [8], [9]. As GCMMA and BFGS clearly showed bad results compared to the remaining six methods, these are not shown in the figure.

The higher the dark column, the more inaccurate is the optimization method. This height is computed as the relative difference between minimum objective function value reached and the most accurate optimization method (lowest objective function value), without regard for CPU time. The higher the gold column, the more time consuming the method. This height is computed as the relative difference between CPU time required and the quickest optimization method, without regard to objective function value.

At first, looking at the robustness of these methods, LGC seems to be the best. Actually, it always led to the researched optimum with an excellent accuracy. This method always led to the minimum objective function value well less than 2.5 %. For average as well as maximum value, LGC seems to be the most robust and accurate.

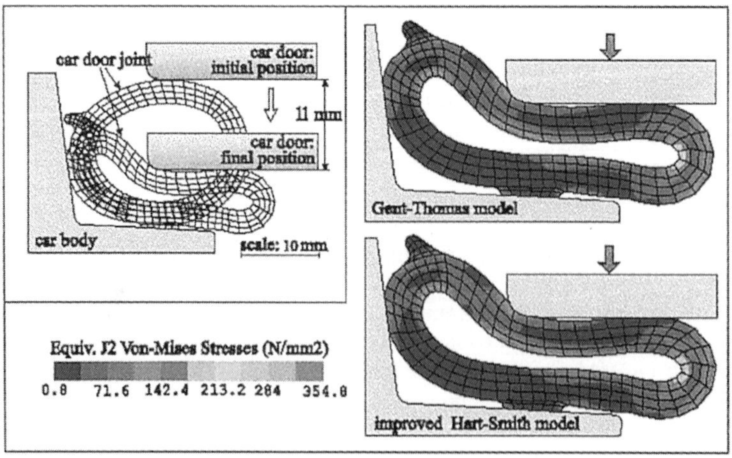

Figure 9 Optimum models comparison in a car door joint crushing test

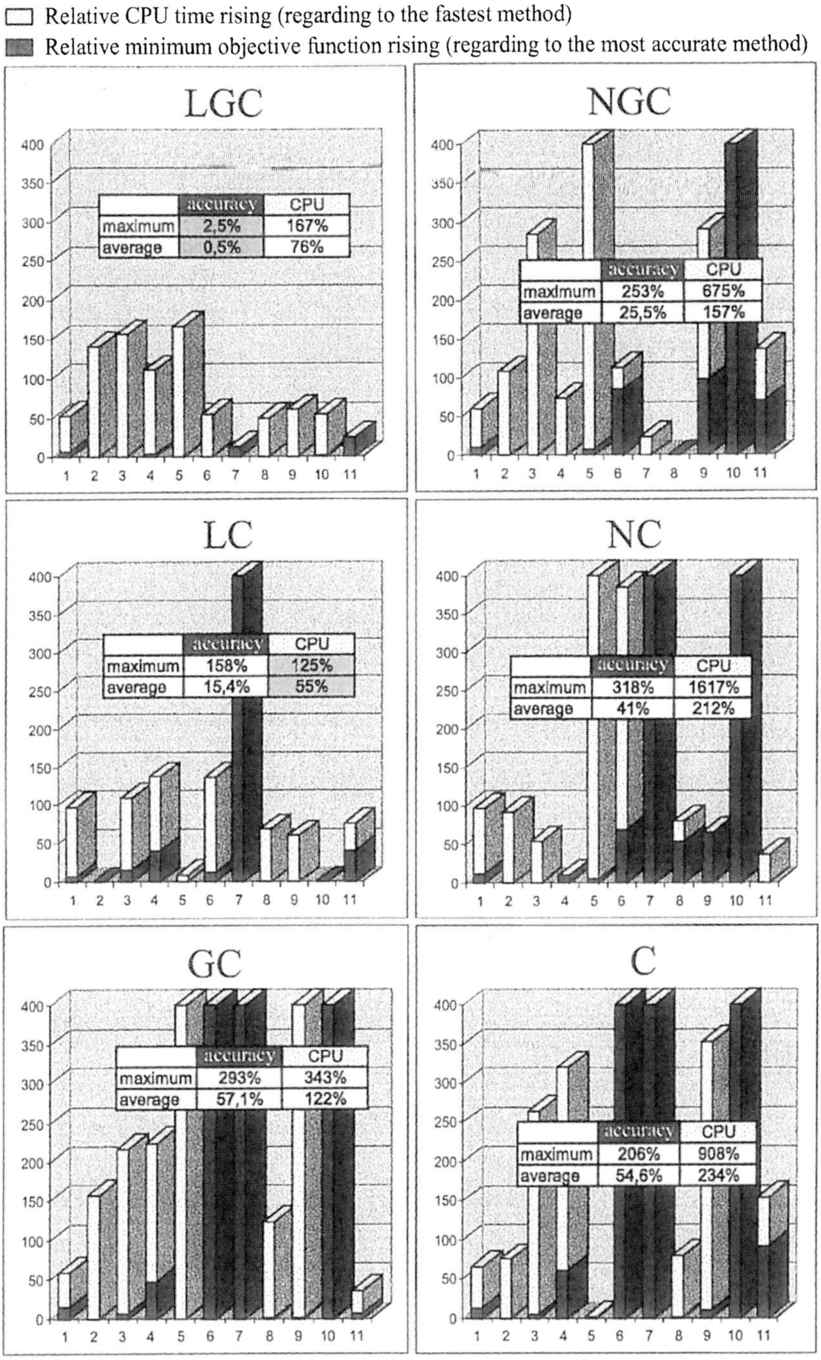

Figure 10 Optimization methods comparison

Unfortunately, this method is not the quickest, but quite takes the second place. The LC is the fastest, for average value and maximum time consuming. However, this last method seems slightly less accurate and robust than the previous one (average accuracy value increase from 0.5 to 15 % and example 7 shows non convergence of the method). Actually, using LGC instead of LC leads to an increase of 14% of CPU time for relative decrease of the optimum objective function of 13 %. Both these methods seems efficient and shows interesting properties. Nevertheless, we think that parameter identification problem can be seen as an investment which will allow us to reach accurate and reliable results in all the future direct simulations that will be computed with the resulting optimum material model. That is why robustness and accuracy seems to be the most important properties of an optimization method dedicated to parameter identification problems. Hence, the most efficient optimization method for considered inverse problems seems to be LGC which will be used by default in future numerical applications.

6. CONCLUSIONS

The coupling of efficient optimization algorithms with a large strain finite element code allow us to consider some new problems. Parameter identification is an inverse problem consisting in minimizing a least-square gap between an experimental result and its FEM simulation in order to find the most appropriated material behavior model. This problem can be formulated as an optimization problem consisting in minimizing an implicit non convex and strongly non linear objective function under boundary and linear constraints.

We proposed several optimization methods (classical or new) and some combinations of these methods in order to find a robust and accurate method. We described several numerical applications showing that the proposed methods are both efficient and useful in a wide variety of parameter identification applications.

We compared the eight proposed optimization methods in term of robust-ness, accuracy and time consumption, based on statistics from eleven numerical applications. It appears that LGC method is the most efficient method for parameter identification problems.

References

[1] C. FLEURY. Sequential convex programming for structural optimization problems. In Rovaznus, editor, *Optimization of Large Structural Systems*,1993.

[2] C. FLEURY. *Optimisation des Structures*. Centrale des Cours del'A.E.E.S., Université deLiège, 1994.

[3] A. GAVRUS, E. MASSONI, and J.L. CHENOT. A finite element model for the simulation of the torsion and torsion-tension tests. *Computer Methodsin Applied Mechanics and Engineering*, 103:417–434, 1993.

[4] A. GAVRUS, E. MASSONI, and J.L. CHENOT. An inverse finite element analysis applied to viscoplastic parameter identification. In John Wiley and Sons, editors, Second *ECCOMAS Conference on Numerical Methods in Engineering*, Paris, September 1996.

[5] J.C. GELIN and O. GHOUATI. Une m´éthode d'identification inverse des paramètres matériels pour les comportements non-linéaires. *Revue européenne des éléments finis*, 4(4):463–485, 1995.

[6] O. GHOUATI. *Identification et Modélisation Numérique Directe et Inversedu comportement Viscoplastique des Alliages d'Aluminium*. PhD thesis, U.F.R. des Sciences et Techniques de l'Université de Franche-Comté, July 1994.

[7] J.P. KLEINERMANN. Modélisation du cintrage de composants aéronautiques par creep-forming et superplasticité. Master's thesis, Université de Liège, June 1996.

[8] J.P. KLEINERMANN. Optimisation des procédés de mise à forme des matériaux par calcul inverse et calcul de sensibilite. F.R.I.A. activity report, Augustus 1997.

[9] J.P. KLEINERMANN. Optimisation des procédés de mise à forme des matériaux par calcul inverse. F.R.I.A. activity report, Augustus 1999.

[10] R. MAHNKEN and E. STEIN. A unified approach for parameter identification of inelastic material models in the frame of the finite element method. *Computer Methods in Applied Mechanics and Engineering*, 136:225–258, 1996.

[11] R. MAHNKEN and E. STEIN. Parameter identification for finite de-formation elasto-plasticity in principal directions. *Computer Methods in Applied Mechanics and Engineering*, (147):17–39, 1997.

[12] A.J. MORRIS. *Foundations of Structural Optimization: A Unified Approach.* 1982.

[13] D.M. NORRIS, J.R.B. MORRAN, J.K. SCUDDE, and QUINONES D.F.A computer simulation of the tension test. *Journal of Mechanical Physicsof Solids*, 26:1–19, 1978.

[14] J.P. PONTHOT. *Mode d'emploi de la version pilote de METAFOR, module de calcul en grandes deformations*. LTAS-Thermomecanique et milieux continus, Université de Liège, http://www.ulg.ac.be/ltas-mct, may 1992.

[15] J.P. PONTHOT, D. ROZENWALD, and CHAMBARD A. Modélisation de l'ecrasement d'un joint de portière à l'aide du logiciel metafor. Technical report, LTAS-Thermomécanique et Milieux Continus, Université de Liège, february 1993.

[16] D. ROZENWALD. *Modélisation Thermomécanique des Grandes Déformations. Application aux Problèmes de Mise à Forme des Métaux, des Elastonères et des Structures Mixtes Métal-Elastomère*. PhD thesis, Université de Liège, november 1996.

[17] J.C. SIMO. A framework for finite strain elastoplasticity based on maximum plastic dissipation and the multiplicative decomposition. *Compute Method in Applied Mechanics and Engineering*, 66:199–219, 1988.

[18] K. SVANBERG. The method of moving asymptotes. a new method for structural optimization. *International Journal for Numerical Methods in Engineering*, 24:359–373, 1987.

[19] K. SVANBERG. A globally convergent version of mma without line-search. In N. Olhoff and G.I.N. Rovaznus, editors, *Proceedings of the First World Congress of Structural and Multidisciplinary Optimization, ISSMO*, pages 9–16, Elsevier Science Ltd., Oxford, may 1995.

[20] K. SVANBERG. Non-mixed second derivatives in mma. Technical report, Dept. of Physics, Lund University Sweden, May 1995.

[21] W.H. ZHANG and C. FLEURY. Two-point based sequential convex approximations for structural optimization. In E. Dick M. Hogge, editor,*3ème Congrès National Belge de Mécanique Théorique et Appliquée*, Université de Liège, may 1994.

[22] C. ZILLOBER. A globally convergent version of the method of moving asymptotes. *Structural Optimization*, 6:166–174, 1993.

A ONE STEP PREDICTION ERROR APPROACH TO THE IDENTIFICATION OF VISCOPLASTIC MATERIAL MODELS

Ola Wall*

Division of Solid Mechanics
Lund Institute of Technology
Box 118, SE-221 00 Lund, Sweden

Jan Holst

Division of Mathematical Statistics
Lund Institute of Technology
Box 118, SE-221 00 Lund, Sweden

Abstract A new approach to the problem of estimating parameter in material models is presented. The approach is based on a state space representation of the constitutive equations and one step predictions. The differences between one-step predictions and the corresponding measurements are used to design generic one-step prediction error estimators, and in particular, the maximum likelihood method is presented. The one-step predictions are computed through extended Kalman filtering. Consequences of using a time dependent model with least squares regression are analysed. It is shown that if the residuals are a sequence of stochastic variables, correlated with the regressors, the parameter estimates may be biased.

A Monte Carlo study shows that the model parameters of a Norton viscoplastic model are estimated with up to 40% higher precision with the new approach as compared to standard least squares regression. An analysis of the residuals clearly shows that the residuals of the new estimators form an independent sequence of random variables.

Keywords: material models, visco-plasticity, identification

* Supported by the Swedish Research Council for Engineering Sciences. Presently at Kockums AB,

P. Ståhle and K.G. Sundin (eds.), IUTAM Symposium on Field Analyses for Determination of Material Parameters - Experimental and Numerical Aspects, 101–112.

1. Introduction

One consequence of the increasing use of complex constitutive equations is the numerous articles which have been published recently on estimation of parameters in material models , see e.g. Courage et al., 1990, Mahnken and Stein, 1996a; Mahnken and Stein, 1996b, Schwertel and Schinke, 1996 and Senseny and Fossum, 1995; Fossum, 1997; Fossum, 1998. Most of these articles are based on statistical methods, primarily least squares regression, or on methods derived from a minimisation of a cost function of some residuals. The two approaches are similar, especially when quadratic optimisation criteria are used. One advantage of the statistical interpretation of the problem is that it provides an estimate of the goodness of fit in terms of the parameter's covariance. The residuals are defined as the differences between some predicted strain values and the corresponding measured strain values, assuming the stresses, or at least the external loads, are known. Naturally, it is also possible to perform the identification with displacement control, predicting the stresses instead of the strains.

The papers of Mahnken and Stein, 1996a, Schwertel and Schinke, 1996 and Fossum, 1997; Fossum, 1998; Senseny and Fossum, 1995 all treated homogeneous uniaxial conditions. Courage et al., 1990 and Mahnken and Stein, 1996b incorporated finite element calculations in the identification process in order to handle multi dimensional problems. The work of Courage et al., 1990 treat a hyper elastic material while the others consider time dependent materials.

The majority of all estimation techniques relies on some assumptions about the distribution of the data or the parameters. Deviation from this assumed distribution may result in biased estimators, i.e. the mathematical expectation of the parameter estimates is not equal to the true parameter values.

In this work, consequences of using a time dependent model with least squares regression are analysed. It will be shown that when the residuals are a sequence of stochastic variables, correlated with the regressors, the parameter estimates may be biased. An in depth analysis of this problem and ways to work around it is found in the book by Gallant, 1987.

A new approach will be presented, based on a state space representation of the constitutive equations and one step predictions. It is inspired by results from time series analysis, see Madsen and Holst, 1999, signal processing, see Jazwinski, 1970, and automatic control and identification, see Ljung, 1998. The basic idea is to compute one step predictions of the strain values. The integration process is extended to perform a state update after each integration step. This process is called state filtering, Jazwinski, 1970, and consist of repeating the following three steps. First, from the current state, the constitutive equations are integrated to a new time instant to give the total strain. The prediction error is obtained by comparing the calculated value with the measured quantity. Second, the statistics of the predicted variable are computed using the dynamics of the constitutive model. Third, update the current state of the constitutive model. The updating is based both the predicted and measured values and their statistics.

In practice, the solution of the state filtering problem requires a few approximations. Here, extended Kalman filtering, as presented by Wall and Holst, 2001, will be used. It is able to produce residuals with approximately white noise properties. From the residuals, generic one-step prediction error estimators are derived. In particular, the maximum likelihood method will be presented.

The new approach will be compared with the least squares regression method by Monte Carlo simulations.

Notations

A few notations will be introduced first. The measurements of strains are taken at discrete time instances t_k, $k \in \{0,\ 1,\ \cdots,\ N\}$. A subscript will denote the time t_k of observation, $\boldsymbol{\epsilon}(t_k) = \boldsymbol{\epsilon}_k$. A subset of the observations is $\boldsymbol{\epsilon}^j = \{\boldsymbol{\epsilon}_j,\ \boldsymbol{\epsilon}_{j-1},\ \cdots,\ \boldsymbol{\epsilon}_1,\ \boldsymbol{\epsilon}_0\}$ and $\boldsymbol{\epsilon}^N$ denotes the entire measurement set. The expectation of a variable will be indicated by a hat, $\hat{a} = E[a]$. Conditioning of a on b will be denoted $(a|b)$. The conditional expectation of $\boldsymbol{\epsilon}_k$ on $\boldsymbol{\epsilon}^{k-1}$ is denoted $\hat{\boldsymbol{\epsilon}}_{k|k-1}$.

2. Least squares regression

When using time dependent constitutive equations, the mathematical formulation consists of differential equations. After employing a numerical method to integrate these equations from time t_{k-1} to time t_k, they can be written as

$$\epsilon_k = F(\sigma_k, \epsilon_{k-1}, \boldsymbol{\theta}) \tag{1}$$

with ϵ_k denoting the total strain at time t_k, σ is the stress and $\boldsymbol{\theta}$ is a vector containing the parameters of the material model. As ϵ_k is the variable which is calculated from the model, it it also called the dependent variable.

In order to illustrate how disturbances may effect a least squares regression estimator, it is assumed that the model is studied in an interval where a linearisation is appropriate. A suitable representation for regression is then

$$\epsilon_k = [\sigma_k,\ \epsilon_{k-1}][\theta_1,\ \theta_2]^T + w_k \tag{2}$$

where the disturbance w_k has been introduced to represent both model and measurement errors. The strain recordings, the stresses and the disturbances are collected in matrices

$$\boldsymbol{\epsilon} = \begin{bmatrix} \epsilon_1 \\ \epsilon_2 \\ \vdots \\ \epsilon_N \end{bmatrix} \quad \boldsymbol{R} = \begin{bmatrix} \sigma_1 & \epsilon_0 \\ \sigma_2 & \epsilon_1 \\ \vdots & \vdots \\ \sigma_N & \epsilon_{N-1} \end{bmatrix} \quad \boldsymbol{W} = \begin{bmatrix} w_1 \\ w_2 \\ \vdots \\ w_N \end{bmatrix}. \tag{3}$$

Here $\boldsymbol{R}$ denote the regressors. The least squares estimator is then

$$\hat{\boldsymbol{\theta}} = \boldsymbol{\theta}_0 + \left(\frac{1}{N}\boldsymbol{R}^T\boldsymbol{R}\right)^{-1}\left(\frac{1}{N}\boldsymbol{R}^T\boldsymbol{W}\right) \tag{4}$$

with $\boldsymbol{\theta}_0$ denoting the true parameter value, i.e. the parameters which are generating the measured strain values. The estimator is unbiased if the expected value

$$E\left[\left(\frac{1}{N}\boldsymbol{R}^T\boldsymbol{R}\right)^{-1}\left(\frac{1}{N}\boldsymbol{R}^T\boldsymbol{W}\right)\right] = 0. \tag{5}$$

This holds if either the residuals w_k are a sequence of independent stochastic variables or if the regressors are uncorrelated with the residuals. However, as the value of strain at time t_k depends on ϵ_{k-1} in time dependent models, the residual at t_k depends on ϵ_{k-1}. The expectation expressed in equation (5) is therefore not equal to zero and the parameter estimate will therefore be biased.

For problems with dependent residuals with stationary stochastic properties, Gallant, 1987 suggest a transformation of the variables to avoid the bias problem.

3. One step predictions

In order to account for model uncertainties and measurement noise, the constitutive equations will be written as a system of stochastic differential equation on state space form

$$\begin{aligned} d\boldsymbol{\epsilon}^{ie} &= f\left(\boldsymbol{\sigma}, \boldsymbol{\epsilon}^{ie}, \boldsymbol{\theta}, t\right)\, dt + d\boldsymbol{w} \qquad \boldsymbol{\epsilon}^{ie}(t_0) \in \mathcal{N}(\bar{\boldsymbol{\epsilon}}^{ie}_{t_0}, \boldsymbol{P}_{t_0}), \\ \epsilon_k &= \boldsymbol{M}\boldsymbol{\epsilon}^{ie}(t_k) + \boldsymbol{C}\boldsymbol{\sigma}(t_k) + \boldsymbol{v}_k. \end{aligned} \tag{6}$$

Here $\boldsymbol{\epsilon}^{ie}$ denotes inelastic strain, $\boldsymbol{\epsilon}$ the total strain, $\boldsymbol{\sigma}$ the stress and $\boldsymbol{C}$ is the flexibility matrix. The formulation reflects the facts that the evolution of plastic strain is a continuous process and the measurements are taken at discrete time instances t_k. The state variables are written as vector quantities in order to provide a more general framework for the present approach. The plastic strain, $\boldsymbol{\epsilon}^{ie}$, can include several components of plastic strain as well as other hardening variables. This requires multiplication of a gain matrix, $\boldsymbol{M}$, in the calculation of the total strain. The gain matrix may also be useful when treating multi-dimensional problems. Model uncertainties are accounted for by the process noise increments $d\boldsymbol{w}$ with $\boldsymbol{w}$ being a Wiener process. Measurement noise is modelled by the Gaussian process $\boldsymbol{v}$. The noise properties are defined by the covariances $\mathrm{cov}[\boldsymbol{w}_t, \boldsymbol{w}_\tau] = \boldsymbol{Q}(\min(t,\tau))\min(t,\tau)$ and $\mathrm{cov}[\boldsymbol{v}_k, \boldsymbol{v}_k] = \boldsymbol{R}_0$.

One step predictions of the strain, $E[\boldsymbol{\epsilon}_k | By_{k-1}] = \hat{\boldsymbol{\epsilon}}_{k|k-1}$ are calculated through extended Kalman filtering. An algorithm has been derived by Wall and Holst, 2001. The Kalman filter provides means for updating the prediction of strain with feedback information from the measurements, i.e.

$$\hat{\boldsymbol{\epsilon}}^{ie}_{t_{k+1}|t_{k+1}} = \hat{\boldsymbol{\epsilon}}^{ie}_{t_{k+1}|t_k} + \boldsymbol{K}_{k+1}\left(\boldsymbol{\epsilon}_{k+1} - \hat{\boldsymbol{\epsilon}}_{k+1|k}\right). \tag{7}$$

For the details of computing K_{k+1}, cf. Wall and Holst, 2001.

3.1 Estimators

The one step prediction error, corresponding to the one step predictor is

$$\tilde{\boldsymbol{\epsilon}}_k(\boldsymbol{\theta}) = \boldsymbol{\epsilon}_k - \hat{\boldsymbol{\epsilon}}_{k|k-1}. \tag{8}$$

A generic estimator is obtained by minimisation of a cost function $V(\boldsymbol{\theta}, \boldsymbol{\epsilon}^N)$, defined as

$$V(\boldsymbol{\theta}, \boldsymbol{\epsilon}^N) = \frac{1}{N}\sum_{k=1}^{N} l(\tilde{\boldsymbol{\epsilon}}_k(\boldsymbol{\theta}), \boldsymbol{\theta}), \tag{9}$$

where $l(\cdot)$ is a scalar valued, positive function. A generic estimator is

$$\hat{\boldsymbol{\theta}} = \arg\min_{\boldsymbol{\theta}} V(\boldsymbol{\theta}, \boldsymbol{\epsilon}^N). \tag{10}$$

Under a few regularity assumptions, see Wall and Holst, 2001, the estimate will converge to the true parameter value. According to a central limit theorem, the parameter estimate is a random variable with
$\sqrt{N}\left(\hat{\boldsymbol{\theta}} - \boldsymbol{\theta}_0\right) \to \mathcal{N}(0, \boldsymbol{P}_{\boldsymbol{\theta}})$. The covariance is $\frac{1}{N}\boldsymbol{P}_{\boldsymbol{\theta}} = \mathrm{cov}[\hat{\boldsymbol{\theta}}]$, with $\boldsymbol{P}_{\boldsymbol{\theta}}$ given by

$$\boldsymbol{P}_{\boldsymbol{\theta}} = \left(E\left[\frac{\partial^2 V(\hat{\boldsymbol{\theta}})}{\partial \boldsymbol{\theta}^2}\right]\right)^{-1} NE\left[\left(\frac{\partial V(\hat{\boldsymbol{\theta}})}{\partial \boldsymbol{\theta}}\right)\left(\frac{\partial V(\hat{\boldsymbol{\theta}})}{\partial \boldsymbol{\theta}}\right)^T\right]\left(E\left[\frac{\partial^2 V(\hat{\boldsymbol{\theta}})}{\partial \boldsymbol{\theta}^2}\right]\right)^{-1}. \tag{11}$$

3.1.1 The conditional least squares estimator. Setting the function $l(\cdot)$ in (9) equal to $l(\cdot) = |\cdot|^2$ results in the conditional least squares estimator. The parameter covariance can be calculated from equation (11).

3.1.2 The maximum likelihood estimator. The maximum likelihood method is based on formal statistical inference and is a conceptually different approach from minimising prediction errors. The likelihood function is the probability of obtaining the measurements $p(\epsilon^N;\boldsymbol{\theta})$. The likelihood function is

$$\mathcal{L}(\boldsymbol{\theta},\epsilon^N) = p(\epsilon^N;\boldsymbol{\theta}) = \prod_{k=1}^{N} p(\epsilon_k|\epsilon^{k-1};\boldsymbol{\theta}), \tag{12}$$

where successive applications of the rule $p(a,b) = p(a|b)p(b)$ have been used. The conditional densities $p(\epsilon_k|\epsilon^{k-1};\boldsymbol{\theta})$ are recognised as the densities of the residuals,

$$p(\epsilon_k|\epsilon^{k-1};\boldsymbol{\theta}) = p(\tilde{\epsilon}_k(\boldsymbol{\theta})|\epsilon^{k-1};\boldsymbol{\theta}) = p(\tilde{\epsilon}_k(\boldsymbol{\theta});\boldsymbol{\theta}). \tag{13}$$

The maximum likelihood estimate is obtained by maximising the likelihood function (12), or equivalently, minimising the negative logarithm likelihood, $\hat{\boldsymbol{\theta}}_{ML} = \arg\max_{\boldsymbol{\theta}} \mathcal{L}(\boldsymbol{\theta},\epsilon^N) = \arg\min_{\boldsymbol{\theta}} \left(-\log\mathcal{L}(\boldsymbol{\theta},\epsilon^N)\right)$.

When the residuals are assumed to be stationary Gaussian with covariance matrix $\boldsymbol{R}$, the density function of the residuals is

$$p(\tilde{\epsilon}_k,\boldsymbol{\theta}) = \frac{1}{(2\pi)^{m_\epsilon/2}\sqrt{\det\boldsymbol{R}}}\exp\left(-\frac{1}{2}\tilde{\epsilon}_k^T(\boldsymbol{\theta})\boldsymbol{R}^{-1}\tilde{\epsilon}_k(\boldsymbol{\theta})\right), \tag{14}$$

where m_ϵ is the dimension of $\tilde{\epsilon}_k$. By using the asymptotic relation $1/N\sum_k^N \tilde{\epsilon}_k\tilde{\epsilon}_k^T \to \boldsymbol{R}$, the log likelihood function can be rewritten

$$-\log\mathcal{L} = \frac{N}{2}\log\det\left(\frac{1}{N}\sum_{k=1}^{N}\tilde{\epsilon}_k(\boldsymbol{\theta})\tilde{\epsilon}_k^T(\boldsymbol{\theta})\right) + \frac{Nm_\epsilon}{2}(1+\log 2\pi) \tag{15}$$

when $\boldsymbol{R}$ is unknown.The parameter's covariance is calculated as $\operatorname{cov}[\hat{\boldsymbol{\theta}}_{ML}] = \mathcal{M}^{-1}|_{\boldsymbol{\theta}=\hat{\boldsymbol{\theta}}}$ with the Fisher information matrix $\mathcal{M}$ defined by

$$\begin{aligned}\mathcal{M} &= E\left[\left(\frac{\partial}{\partial\boldsymbol{\theta}}\log p(\epsilon^N;\boldsymbol{\theta})\right)\left(\frac{\partial}{\partial\boldsymbol{\theta}}\log p(\epsilon^N;\boldsymbol{\theta})\right)^T\right]\Bigg|_{\boldsymbol{\theta}=\boldsymbol{\theta}_0} \\ &= -E\left[\frac{\partial^2}{\partial\boldsymbol{\theta}^2}\log p(\epsilon^N;\boldsymbol{\theta})\right]\Bigg|_{\boldsymbol{\theta}=\boldsymbol{\theta}_0}\end{aligned} \tag{16}$$

The Cramer-Rao inequality states that the covariance of a parameter estimator is $\operatorname{cov}[\hat{\boldsymbol{\theta}}] \geq \mathcal{M}^{-1}$, Ljung, 1998. If the covariance takes the lower bound of the Cramer-Rao inequality, the estimator is said to be efficient. If there exists an unbiased efficient estimator, it is the maximum likelihood estimator, cf. Wall and Holst, 2001. For the stationary Gaussian case, the Fisher information can be computed through equation (15).

4. Results

The theory presented above has been evaluated with simulations of a generalised Norton model suitable for transient loading problems, see Wall, 1998. The primary goal has been to compare the one-step prediction approach to standard regression. A block diagram of the evaluation setup is shown in figure 1.

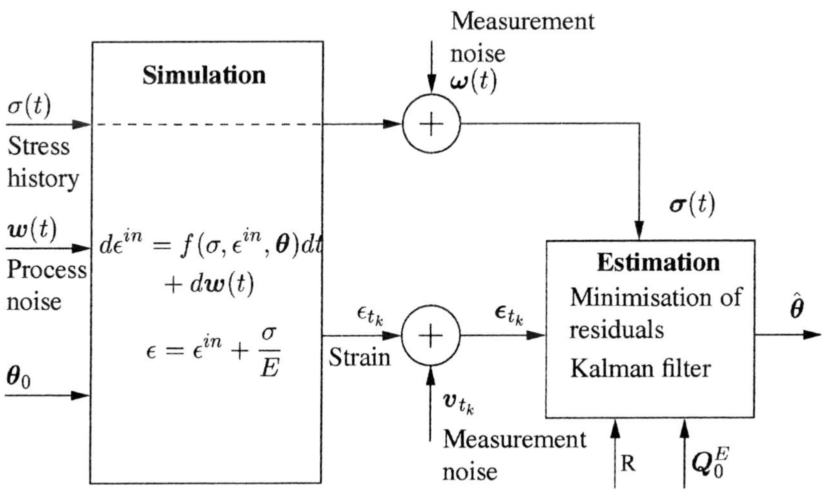

Figure 1. The setup for evaluating the estimator. The estimated parameters $\hat{\theta}$ should be equal to the true parameters θ_0 which is used for generation of the data.

The statistics of the added noise processes is given by $(w_{t_k} - w_{t_{k-1}}) \in \mathcal{N}(0, Q_0)$, $v_k \in \mathcal{N}(0, R_0)$ and $(\omega_{t_k} - \omega_{t_{k-1}}) \in \mathcal{N}(0, \Omega)$, as the Wiener processes produces Gaussian increments between two time instances. The outputs from the simulations are then used as inputs to the estimator. The parameter estimator also requires knowledge of the noise properties, which are given by Q_0^E and R_0^E. The result is the estimate of the parameters, $\hat{\theta}$. The estimator was found to be sensitive to high frequency noise in the stress signal and therefore, it was pre-filtered by a linear phase low pass filter before estimating the parameters.

The solution of the estimation problem is obtained optimisation. The Gauss-Newton method has been used successfully here, cf. Fletcher, 1987 Both the simulation and the estimation require numerical integration. An explicit fifth order Runge-Kutta method was used with a convergence criterion with an absolute tolerance of 10^{-7} for the inelastic strain.

The uniaxial constitutive equation for the Norton model is

$$\dot{\epsilon}^{ie} = f_\sigma\left(\sigma, \epsilon^{ie}, \boldsymbol{\theta}\right) = \begin{cases} \left(\frac{\sigma - \left(h(\epsilon^{ie})^n + \sigma_y\right)}{\eta}\right)^\gamma & \text{if } \sigma > \left(h(\epsilon^{ie})^n + \sigma_y\right), \\ 0 & \text{if } \sigma \le \left(h(\epsilon^{ie})^n + \sigma_y\right) \end{cases} \tag{17}$$

Isotropic hardening is controlled by the hardening parameters h and n. The time dependent, viscous, properties are governed by γ and η. The parameters were chosen to represent a structural steel, with Young's modulus E = 206GPa, an initial yield stress σ_y = 413MPa and the hardening parameters $n = 0.8$ and h = 500MPa. The viscous parameters were $\gamma_0 = 2$ and η_0 = 10MPa. Curves of stress-strain without any added noise are shown in figure 2. It can be seen that the model responds nearly rate independent at the lower strain rates. The strain hardening parameters h and n can therefore be found by ordinary least squares regression from the low rate response, leaving only the viscous part to be identified by the conditional least squares method. The benefits of splitting the problem into one rate independent, static, part and one viscous part are faster and more robust computations. Hence the unknown parameters are $\boldsymbol{\theta} = [\gamma \;\; \eta]$.

Ljung, 1998 discusses the choice of suitable sampling time, and suggests $T = t_k - t_{k-1} = -\frac{1}{2}\left(\text{minimum eigenvalue of }\left(\frac{\partial f}{\partial \epsilon^{ie}}\right)\right)^{-1}$. At the maximum strain rate, which is expected to be at most 10^3 s^{-1} and the approximate inelastic strain is 0.1, the sampling time should be less than $T = (8000)^{-1}$s. The model was simulated at three different strain rates, $\dot{\epsilon} = [10^1, \;\; 10^2, \;\; 10^3]$ s^{-1}.

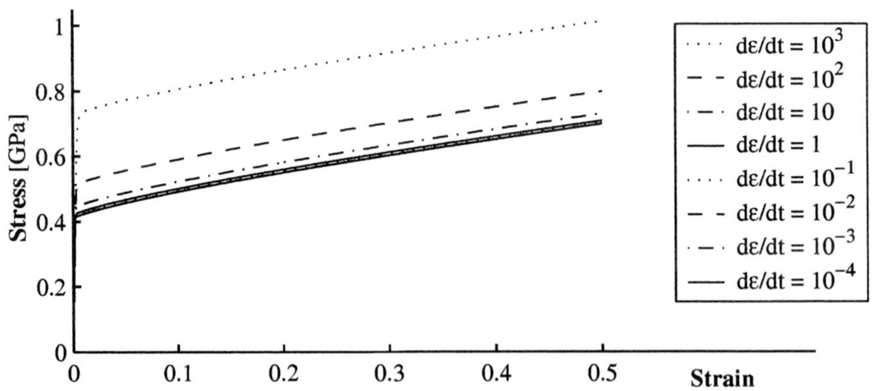

Figure 2. Simulated stress-strain for the generalised Norton model for transient loading.

At the different strain rates, the sampling interval was chosen as $T = [10^{-4}, 10^{-5}, 10^{-6}]$s.

A total of 1500 points which were used for each parameter estimate. 200 different parameter estimates were calculated. The mean values and their statistics are shown in table 1. It can be concluded that when fairly correct noise statistics are supplied to the estimator, the conditional least squares estimator has a higher precision than the least squares regression estimator. For the latter, the precision is between 40% to 60% worse. Using the wrong noise statistics in the conditional least squares estimator, as in cases 1 and 2, results in decreasing precision. When noise is added to the input signal, cases 9 and 10, these differences become less pronounced. In the cases of high process noise, cases 7, 8, 15 and 16, the model represents an inaccurate description of the material to be identified. In these cases, the conditional least squares estimator sometimes, about 10% of the investigated samples, failed to find any parameter estimates whereas the regression estimator always succeeded.

A sample of the residuals from one estimation is shown in figure 4 with noise statistics corresponding to case 5. The conditional least squares estimator produces uncorrelated Gaussian random residuals. The residuals from the least squares regression estimator, on the other hand, illustrate that the the regression estimator cannot deal with measurements which drift away from the model outputs. Further, the residuals deviates slightly from the Gaussian distribution but the distribution is, however, symmetric. The parameter estimates is therefore probably rather reliable.

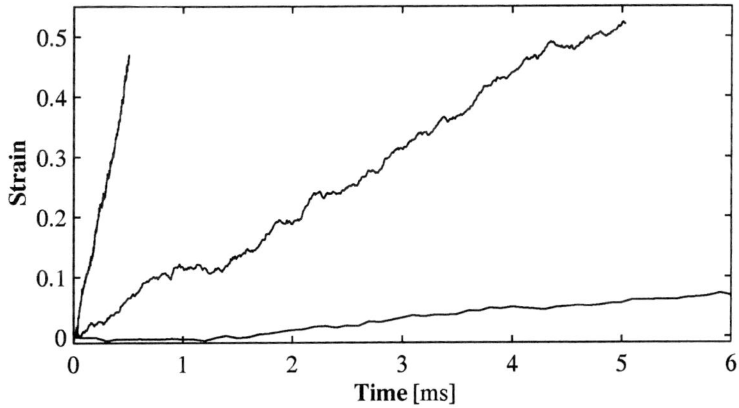

Figure 3. *A typical strain output from the simulation of the generalised Norton model for transient loading. The curves represent loading at three different speeds, and only one tenth of the slowest curve is shown. The noise statistics corresponds to case 5 in table 1.*

No	Noise characteristics			$\hat{\gamma}_{\mathrm{CLS}}$		$\hat{\gamma}_{\mathrm{LSQ}}$		$\hat{\eta}_{\mathrm{CLS}}$		$\hat{\eta}_{\mathrm{LSQ}}$	
	$\sqrt{\boldsymbol{Q}_0^E}$	$\sqrt{\boldsymbol{Q}_0}$	$\sqrt{\boldsymbol{R}_0}$	mean	std	mean	std	mean	std	mean	std
1	0.001	0	0.0025	2.00	0.002	2.00	0.001	9.96	0.03	10.0	0.02
2			0.005	2.00	0.004	2.00	0.002	9.96	0.06	10.0	0.04
3		0.001	0.0025	2.00	0.029	2.00	0.043	10.0	0.37	10.0	0.60
4			0.005	2.00	0.025	2.00	0.040	10.0	0.33	10.0	0.57
5	0.0025	0.0025	0.0025	1.99	0.069	1.99	0.11	9.97	0.86	9.97	1.45
6			0.005	1.99	0.073	1.98	0.11	9.91	0.87	9.80	1.45
7	0.005	0.005	0.0025	1.92	0.27	1.98	0.25	9.89	2.81	9.75	3.34
8			0.005	1.94	0.22	1.94	0.23	9.74	2.38	9.30	2.93
9	0.001	0	0.0025	2.01	0.012	2.00	0.010	10.1	0.19	10.0	0.17
10			0.005	2.00	0.013	2.00	0.010	10.1	0.20	10.0	0.16
11		0.001	0.0025	2.01	0.033	2.00	0.044	10.2	0.44	10.0	0.63
12			0.005	2.01	0.029	2.01	0.041	10.1	0.39	10.1	0.58
13	0.0025	0.0025	0.0025	2.00	0.070	1.98	0.11	10.1	0.89	10.0	1.74
14			0.005	2.00	0.074	1.99	0.11	10.0	0.91	9.83	1.46
15	0.005	0.005	0.0025	1.93	0.26	1.98	0.26	9.89	2.75	9.78	3.35
16			0.005	1.95	0.21	1.95	0.23	9.86	2.30	9.34	2.96

Table 1. The estimates of the parameters for a generalised Norton model for transient loading. The first eight cases were performed with $\boldsymbol{\Omega} = 0$ *and cases 9 to 16 with* $\boldsymbol{\Omega} = (0.02\sigma)^2$. *Each estimation was performed with one time series each at the strain rates* $\dot{\epsilon} = [10,\ 10^2,\ 10^3]\,s^{-1}$.

5. Summary and conclusions

It has been shown how the use of ordinary regression for parameter estimation in dynamic models may result in biased estimates, A one step prediction error approach for the problem of identifying material models has been presented, from which a family of estimators can be derived. The presentation has been accompanied with expressions for the variance-covariance matrices of the estimators. The maximum likelihood estimator is statistically efficient and it offers a flexible framework to handle residuals of any distribution. For Gaussian distributed residuals, the maximum likelihood parameter estimator is similar to the conditional least squares estimates.

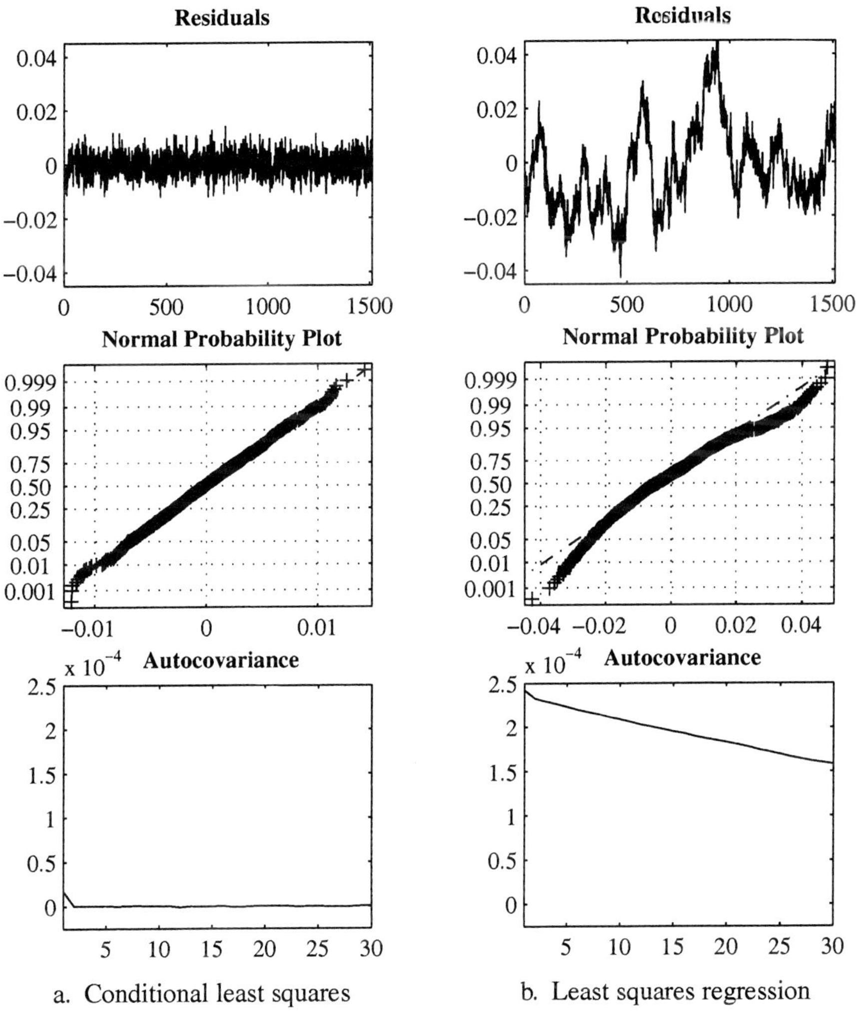

Figure 4. A comparison of the residuals. They were generated with noise statistics corresponding to case 5 in table 1.

A comparison between a traditional least squares regression estimator and the one-step prediction approach showed that the new method produces more accurate estimates and uncorrelated residuals. The precision increased up to 40% when used with a generalised Norton model in transient loading.

The one-step prediction estimators require that the noise statistics is known. The improved precision may be wasted if the wrong noise statistics are supplied to the estimators. In a more advanced implementation, it is possible to estimate the noise statistics simultaneously with the one step predictions, see Madsen and Holst, 1999.

When stress-strain data from a real structural steel were analysed, it turned out that the least squares regression provided more consistent results than the conditional least squares estimator. This was probably caused by inaccurate constitutive equations. The one step prediction approach seems to be sensitive to inaccurate constitutive material models.

Further evaluation of one step prediction error estimators are found in Wall and Holst, 2001, which includes both various levels of excitation as well as models for high temperature creep.

References

Courage, W., Schreurs, P., and Janssen, J. (1990). Estimation of mechanical parameter values of composites with the use of finite element and system identification techniques. *Computers and Structures*, 34(2):231–237.

Fletcher, R. (1987). *Practical methods of optimization*. John Wiley and Sons, 2 edition.

Fossum, A. (1997). Parameter estimation for an internal variable model using nonlinear optimization and analytical/numerical response sensitivities. *Transactions of the ASME, Journal of Engineering Materials and Technology*, 119:337–345.

Fossum, A. (1998). Rate data and material model parameter estimation. *Transactions of the ASME, Journal of Engineering Materials and Technology*, 120:7–12.

Gallant, A. (1987). *Nonlinear statistical models*. John Wiley and Sons, New York.

Jazwinski, A. (1970). *Stochastic processes and filtering theory*. Academic Press, New York.

Ljung, L. (1998). *System identification. Theory for the user*. Prentice Hall, 2 edition.

Madsen, H. and Holst, J. (1999). Modelling non-linear and non-stationary time series. Lecture notes, Department of Mathematical Modelling, Technical University of Denmark, Lyngby, Denmark.

Mahnken, R. and Stein, E. (1996a). Parameter identification for viscoplastic models based on analytical derivatives of a least squares functional and stability investigations. *International Journal of Plasticity*, 12(4):451–479.

Mahnken, R. and Stein, E. (1996b). A unified approach for parameter identification of inelastic material models in the frame of the finite elements method. *Computer methods in applied mechanics and engineering*, 136(3-4):225–258.

Schwertel, J. and Schinke, B. (1996). Automated evaluation of material parameters of viscoplastic constitutive equations. *Transactions of the ASME, Journal of Engineering Materials and Technology*, 118:273–280.

Senseny, P. and Fossum, A. (1995). On testing requirement for viscoplastic constitutive parameter estimation. *Transactions of the ASME, Journal of Engineering Materials and Technology*, 117:151–156.

Wall, O. (1998). Viscoplastic material modelling by thermal activation. Technical report, Department of Solid Mechanics, Lund University, Sweden.

Wall, O. and Holst, J. (2001). Estimation of parameters in viscoplastic and creep material models. *SIAM Journal on Applied Mathematics*, 61:2080–2103.

AUTOMATED IDENTIFICATION AND CALCULATION OF THE PARAMETERS OF NON-LINEAR MATERIAL MODELS

Ralf Mohrmann
Fraunhofer-Institut für Werkstoffmechanik
D-79108 Freiburg, Wöhlerstrasse 11, Germany

Abstract. Material models can be defined as a set of ordinary differential equations or as an analytical function. Usually, material models include model parameters. These model parameters are adjusted with respect to experimental data. A numerical tool, called FitIt [1], for an automated optimisation of model parameters is described. The underlying methods are the Levenberg-Marquardt method and the simple Gradient-method. First and second order derivatives of a quality function with respect to the model parameters are used. The derivatives are calculated numerically; an empirical method for the calculation of these derivatives is established.

1. INTRODUCTION

Usually, the development of complex material models is difficult and time consuming. Stability and convergence are the main aspects for choosing a typical numerical method i.e. for non-linear parameter optimisation or for numerical integration of a set of differential equations. Here, these and other numerical methods are discussed. All methods are included in a proprietary software, called FitIt, which is designed to support the development of material models. The model parameters are adjusted with respect to experimental data. FitIt can be used for models defined as a set of ordinary differential equations or as analytic functions, respectively.

2. THEORY AND NUMERICAL METHODS

The theoretical background and the numerical methods are describes here in order understand the details of the implementation of FitIt. The FitIt user interface hides most of these details from the FitIt user. Therefore, readers who are interested in the application of FitIt can skip the present and the following section.

2.1 Non-linear parameter optimisation

χ^2 is a non-linear scalar function of a parameter vector $\vec{p}$ with the dimension n (cf. [1]. The aim is to find the minimum of χ^2. An iterative method has to be used for this purpose.
A 2nd order Taylor expansion of χ^2 at the current parameter vector $\vec{p}_{cur}$ gives:

P. Ståhle and K.G. Sundin (eds.), IUTAM Symposium on Field Analyses for Determination of Material Parameters - Experimental and Numerical Aspects, 113–122.

$$\chi^2(\vec{p}) = \chi^2(\vec{p}_{cur}) + (\vec{p} - \vec{p}_{cur})\nabla\chi^2(\vec{p}_{cur}) + \\ + \frac{1}{2}(\vec{p} - \vec{p}_{cur})\nabla\nabla\chi^2(\vec{p}_{cur})(\vec{p} - \vec{p}_{cur})$$

Near the minimum the following equation can be employed

$$\vec{p}_{\min} = \vec{p}_{cur} - \underline{\underline{A}}^{-1} \cdot \nabla\chi^2(\vec{p}_{cur}) \text{ with } \underline{\underline{A}} = \nabla\nabla\chi^2(\vec{p}_{cur}).$$

Otherwise, a gradient step can be taken

$$\vec{p}_{next} = \vec{p}_{cur} - const.\ \nabla\chi^2(\vec{p}_{cur}).$$

The Levenberg-Marquardt method combines the two former equations: The diagonal elements of $\underline{\underline{A}}$ are multiplied by a factor $(1+\lambda)$. The modified $\underline{\underline{A}}$ is called $\underline{\underline{A}}'$. The parameter increment $\delta\vec{p}$ can then be calculated by resolving the set of linear equations:

$$\underline{\underline{A}}' \cdot \delta\vec{p} = \nabla\chi^2(\vec{p}_{cur}).$$

The model results $y(x_i, \vec{p})$ and $\vec{y}'(x_i, \vec{y}, \vec{p})$ are either given in closed analytical form, or they are calculated by a numerical integration of a set of ordinary differential equations. For the integration an implicit method is used.

2.2 Calculation of the χ^2 function

The χ^2 function can be calculated with different definitions. The standard definition is

$$\chi_y^2(\vec{p}) = \underbrace{\sum_{i=1}^{M}\left[\frac{y_i - y(x_i, \vec{p})}{s_{y,i}}\right]^2}_{S_1}$$

with the number of data points, M, the value of a data point, y_i, the error in this data point, $s_{y,i}$, and the model result for this data point, y. The second defini tion also takes the derivatives, y', of the variable (with respect to x) into account:

$$\chi_{yy'}^2(\vec{p}) = \sum_{i=1}^{M}\underbrace{\left(\left[\frac{y_i - y(x_i, \vec{p})}{s_{y,i}}\right]^2 + \left[\frac{y_i' - y'(x_i, \vec{p})}{s_{y',i}}\right]^2\right)}_{S_2}$$

with $y_i' = \dfrac{y_{i+1} - y_i}{x_{i+1} - x_i}$ and $s_{y',i} = \dfrac{\sqrt{s_{y,i+1}^2 + s_{y,i}^2}}{x_{i+1} - x_i}$ for $1 \le i < M$

and with $y_M' = y_{M-1}'$ and $s_{y',M} = s_{y',M-1}$.

The third definition uses only derivative information for the calculation of χ^2

$$\chi_{y'}^2(\vec{p}) = \underbrace{\sum_{i=1}^{M} \left[\frac{y_i' - y'(x_i, \vec{p})}{s_{y',i}} \right]^2}_{S_3}.$$

In many cases different types of experiments with various numbers of data points are used within FitIt simultaneously. Therefore, each experiment should be assigned to a class. A useful definition for χ^2 is then

$$\chi^2(\vec{p}) = \sum_{j=1}^{J} \left[\omega_j \sum_{k=1}^{K_j} \left(\frac{1}{K_j} \sum_{i=1}^{M_{j,k}} SUM \right) \right]$$

with the number of classes J, the weight ω_j for each class, the number of Experiments K_j of the class j and the number of data points $M_{j,k}$ of Experiment k of class j. SUM is chosen according to one of the three definitions for S_i given before.

2.3 Integration of differential equations

The integration of differential equations is done numerically with an implicit method (cf. [1]. Implicit integration enables the solution of stiff and non-stiff problems, respectively.

The non-linear set of equations

$$\vec{y}_{n+1} = \vec{y}_n + h \cdot \vec{f}(\vec{y}_{n+1}) \text{ with } \vec{y}' = \vec{f}(\vec{y}, x, \vec{p})$$

is solved iteratively. A first order method is the so called semi-implicit Euler method

$$\vec{y}_{n+1} = \vec{y}_n + h \left[1 - h \frac{\partial \vec{f}}{\partial \vec{y}} \right]^{-1} \vec{f}(\vec{y}_n),$$

which is not guarantied to be stable but is very robust. Higher order implicit methods can result in advantages in computing time and/or accuracy. A Rosenbrock method, which is a generalised Runge-Kutta scheme, worked for any practical problem. This method seeks a solution of the form

$$\vec{y}_{n+1} = \vec{y}_n + \sum_{i=1}^{s} c_i \vec{k}_i ,$$

where the corrections $\vec{k}_i$ are found by solving s linear equations

$$\left(1 - \gamma h \frac{\partial \vec{f}}{\partial \vec{y}}\right) \vec{k}_i = h\vec{f}\left(\vec{y}_n + \sum_{j=1}^{i-1} \alpha_{ij} \vec{k}_j\right) + h \frac{\partial \vec{f}}{\partial \vec{y}} \sum_{j=1}^{i-1} \gamma_{ij} \vec{k}_j, \quad i = 1,\ldots,s.$$

The coefficients γ, c_i, α_{ij} and γ_{ij} are fixed constants, independent of the prob lem. For both methods, the calculation of the Jacobian $\partial \vec{f} / \partial \vec{y}$ is done numerically.

2.4 Calculation of $\underline{\underline{A}}^{-1}$: Singular value decomposition

The inverse of the matrix $\underline{\underline{A}}$ has to be calculated within the Levenberg-Marquardt method (cf. [1]. If the current value of a parameter component p_i does not (significantly) influence the χ^2 result, $\underline{\underline{A}}$ becomes singular and cannot be in verted directly. Therefore, the singular value decomposition (SVD) method is used. First, $\underline{\underline{A}}$ is decomposed with the QR-method

$$\underline{\underline{A}} = \underline{\underline{U}} \cdot \underline{\underline{W}} \cdot \underline{\underline{V}}^T \text{ with } \underline{\underline{U}}^T \cdot \underline{\underline{U}} = 1 \text{ and with } \underline{\underline{V}}^T \cdot \underline{\underline{V}} = 1 .$$

All matrices are square. $\underline{\underline{W}}$ is a diagonal matrix; the components of its inverse denote to singular components. Second, the singular components have to be identified. The singular components are zero; these components near to zero are set to zero in order to enable the calculation of the inverse of $\underline{\underline{A}}$

$$\underline{\underline{A}}^{-1} = \underline{\underline{V}} \cdot \underline{\underline{W}}^{-1} \cdot \underline{\underline{U}}^T .$$

2.5 Numerical calculation of derivatives

Derivatives cannot be calculated numerically with an equation like

$$f'(x) = \frac{f(x+h) - f(x-h)}{2h}$$

without any precautions. Truncation and roundoff errors should be taken into account. In our case, f is the numerical result of a χ^2 function and therefore the finite accuracy of f is an additional limitation for the choice of the step size h.

An empirical recipe was found to work well in practical applications:

The step size h is adjusted such, that $f(x+h)$ and $f(x-h)$ differ in one half of the significant digits. If no h is found to fulfil this criteria within the interval $0 < h < 0{,}1|x|$, the derivative is set to zero. This recipe is used for first and second order derivatives.

3. VERIFICATION OF NUMERICAL METHOD

3.1 Exponential function with finite resolution

This examples should illustrate how the recipes for the calculation of derivatives works. The exponential function e^x was used as a model function and evaluated with finite accuracy at $x = 1$. Figure 1 gives the results of the derivatives for three resolutions calculated with

$$y' = \frac{f(x+h) - f(x)}{h}.$$

Clearly, the truncation errors can be seen for step sizes larger than 10% and roundoff errors are present for small step sizes. In case of intermediate step sizes, the numerically calculated derivative well reproduces the analytical result plotted as the horizontal line in figure 1. The vertical lines in figure 1 indicate the number of equal digits for $f(x)$ and $f(x+h)$; these numbers are derived from figure 2. Comparing the results for a resolution of 6 digits (4 digits) with the vertical line for three (two) equal digits, it is obvious, that the above given recipe works well in this case. Figure 2 gives the difference and the relative difference of $f(x+h)$ and $f(x)$ as a function of the applied step size h, respectively.

3.2 The dependence of the quality function on one parameter

This example discusses the dependence of the quality function χ^2 on the pa rameter Z_p, which was arbitrarily chosen from a practical application using a viscoplastic Chaboche-type model adjusted to a large set of experimental data. The variation of the parameter Z_p, which represents the viscosity of the mate rial, was done with all other parameters fixed to their optimised values.

From figure 3 it can be concluded, that both is possible, the χ^2 function is steep or flat. And, the valley for the minimal χ^2 is narrow. In such a case, it is very difficult to identify the minimum of the χ^2 function with a stochastically working method like the Monte-Carlo method. Figure 4 is an enlargement of figure 3, showing that parabolic approximation of χ^2 is possible near the minimum.

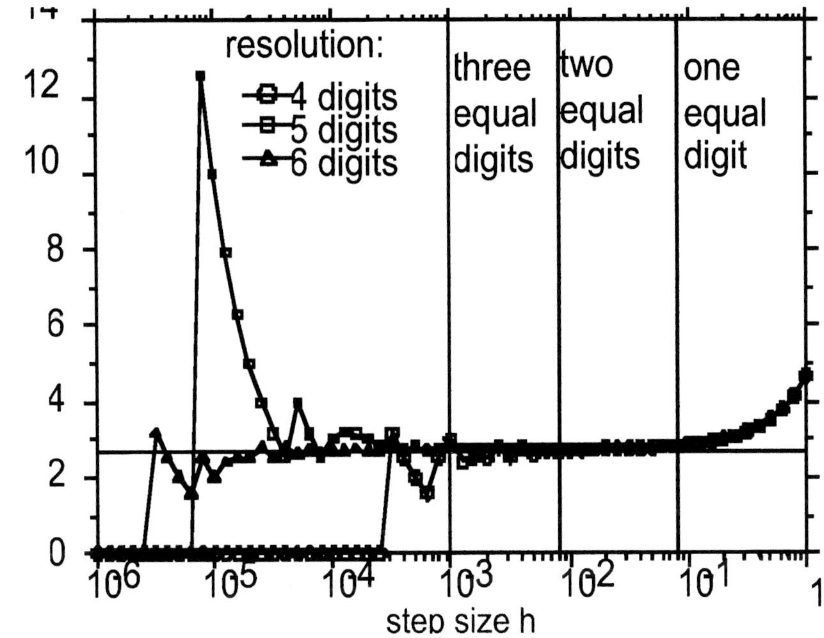

Fig. 1: Numerical evaluation of e^x at $x = 1$ for different resolutions

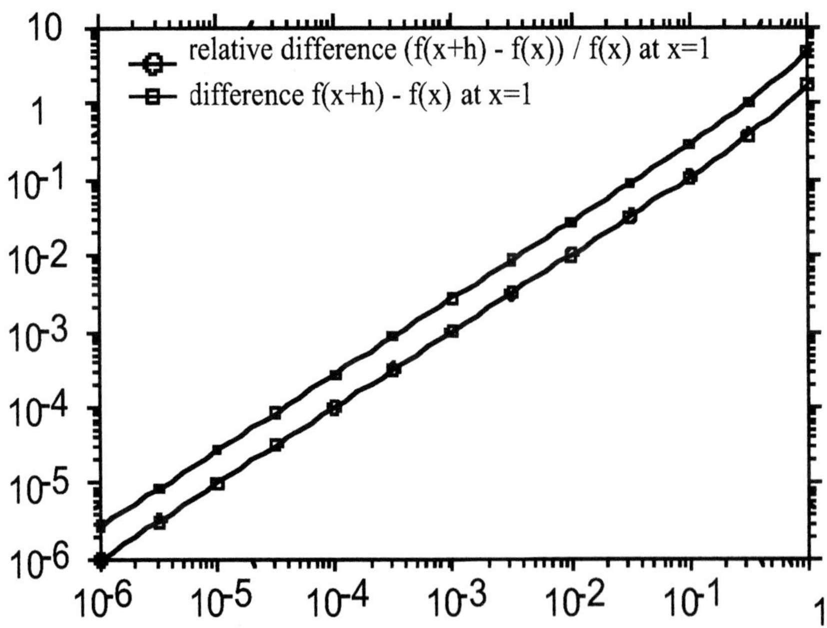

Fig. 2: Differences as a function of the applied step size (same condition than figure 1)

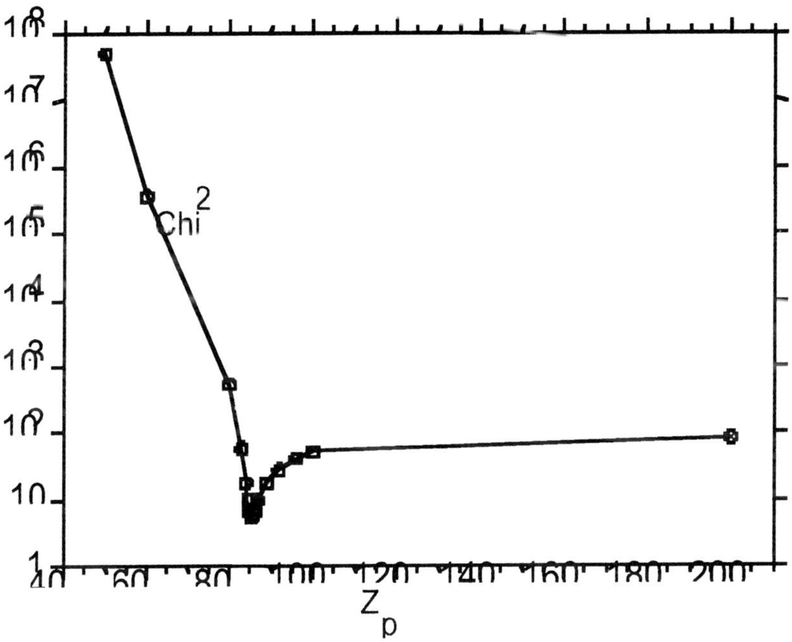

Fig. 3: Dependence of the χ^2 function on one arbitrarily chosen parameter

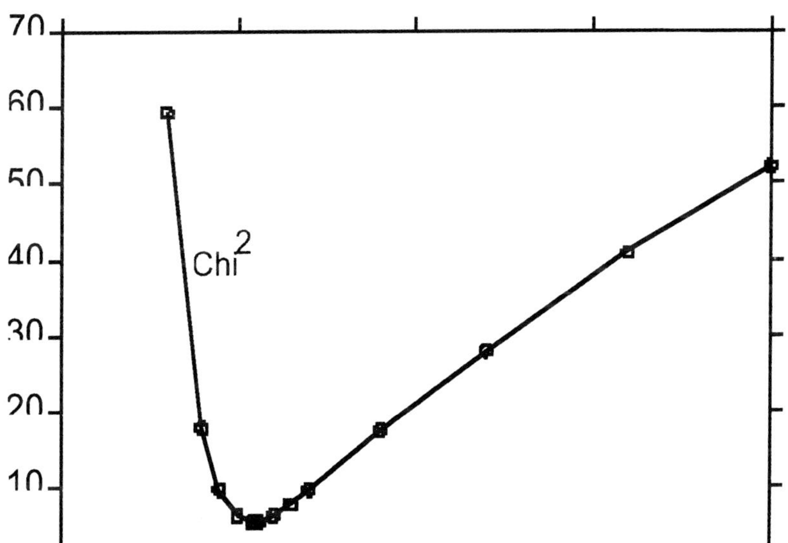

Fig. 4: The χ^2 function near the minimum (enlargement of figure 3)

4. EXAMPLE

As an example, results for a Chaboche type model are described here. The set of differential equations used within FitIt are:

$$\dot{\varepsilon}_{tot} = \dot{\varepsilon}_{pow} + \frac{\dot{\sigma}}{E}$$

$$\dot{\varepsilon}_{pow} = \dot{p}_{pow}\,\mathrm{sgn}\left(\sigma - \alpha_{pow}\right) \text{ with } \dot{p}_{pow} = \left\langle \frac{\left|\sigma - \alpha_{pow}\right|}{Z_{pow}} \right\rangle^{n}$$

$$\alpha_{pow} = \alpha_1 + \alpha_2$$

$$\dot{\alpha}_i = H_i \dot{\varepsilon}_{pow} - D_i \phi_i \alpha_i \dot{p}_{pow} - R\alpha_i \text{ with } i = 1,2$$

$$\phi_1 = \phi_{1,ss} + \left(1 - \phi_{1,ss}\right)\exp\left(-B_1\left|\varepsilon_{pow}\right|\right)$$

$$\phi_2 = \phi_{2,ss} + \left(1 - \phi_{2,ss}\right)\exp\left(-B_2 p_{pow}\right)$$

The model parameters are: E, Z_{pow}, n, $H_{1,2}$, $D_{1,2}$, R, $\phi_{1,ss}$, B_1, $\phi_{2,ss}$, B_2. The parameters were adjusted simultaneously to a large set of experiments including tensile tests, creep tests and cyclic tests. Part of the results are given in the following figures 5-7.

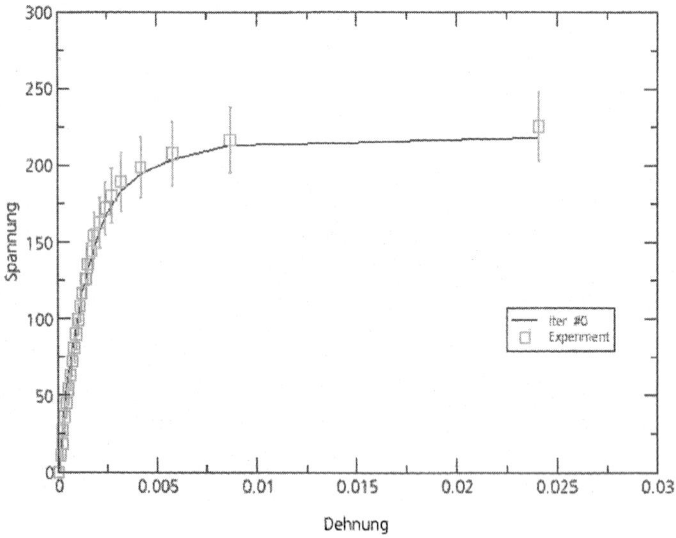

Fig. 5 Tensile test at 625°C Loading rate: 10e-5 1/s Application: stress over total strain units; MPa, s symbols: measurements line: Chaboche Model

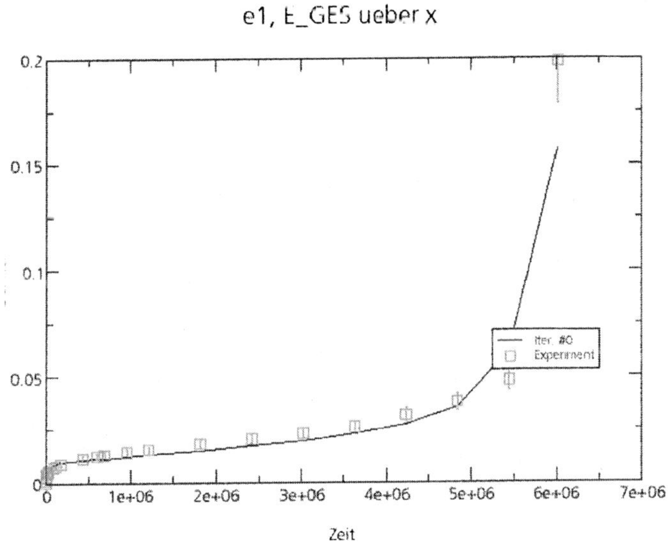

Fig. 6: Creep test at 625°C applied stress: 135 MPa Application: inelastic strain over time units; MPa, s symbols: measurements ine: Chaboche Model

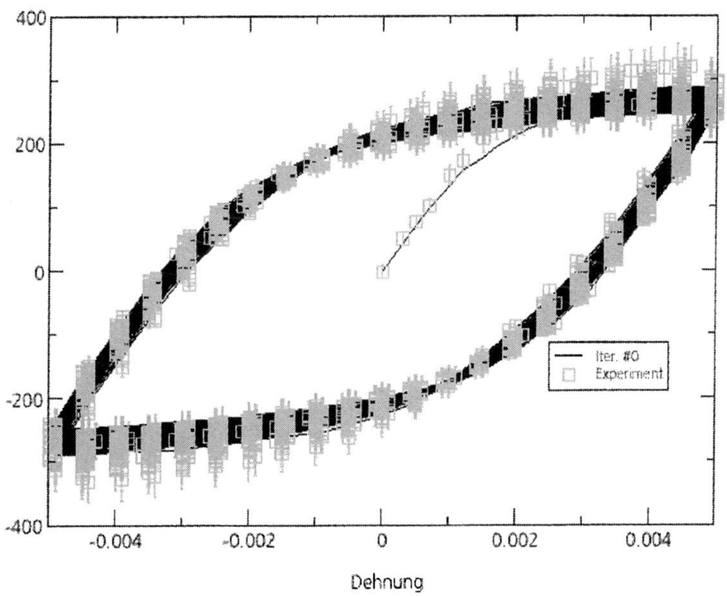

Fig. 7 LCF test at 625°C Loading rate: 10e-3 1/s Application: stress over total strain units; MPa, s symbols: measurements line: Chaboche Model

5. CONCLUSION

The automated optimisation of model parameters is possible for models defined as a set of ordinary differential equations or models defined as an analytical function. Stability and convergence is achieved using the Levenberg-Marquardt method, a suitable choice of the χ^2 function, implicit integration in case of dif ferential equations, the singular value decomposition method and a recipe for the calculation of derivatives, respectively.

Reference

[1] W.H. Press, et al. Numerical Recipes. Cambridge University Press, Cambridge 1986

INVERSION METHOD USING SPECTRAL DECOMPOSITION OF GREEN'S FUNCTION

Muneo Hori
Earthquake Research Institute, University of Tokyo, Yayoi, Tokyo 113-0032, Japan

Abstract An inverse analysis method using the spectral decomposition of Green's function is proposed. For linear inverse problems of identifying inner sources from surface responses, Green's function, which is a solution of the corresponding physical problem, relates the sources to the responses and hence behaves as a compact operator. A suitable inverse operator that maps the responses to the sources can be determined by using the spectral decomposition of Green's function. The proposed method numerically computes the spectral decomposition and determines the inverse operator. It is possible to regard a simple inverse analysis method, which uses a point-wise discretization of Green's function and computes a generalized inverse matrix using the singular-value decomposition, as an approximation to compute the inverse operator. The accuracy, however, could be much lower than the proposed method, because of the point-wise discretization and the less accurate computation of the spectral decomposition. Illustrative examples are solved to demonstrate the usefulness of the proposed inverse analysis method, as well as errors inherent to inverse analysis methods which do not compute the spectral decomposition of Green's function properly.

1. INTRODUCTION

In solid geophysics, solving inverse problems of a semi-infinite body is essentially important; see, for instance, Tarantola (1987), Menke (1989), Fowler (1990) and Dimri (1992); see also Sen and Stoffa (1995). Such problems are the identification of sources located in crust by making use of data measured on ground surfaces. Typical problems are the prediction of dislocation or back slip on a fault plane, the determination of the density distribution and the electromagnetic induction of the crust; see recent articles such as Ranbothen and Pratt (1997), Hornby *et al.* (1999) and Yao and Roberts (1999). Although these problems have distinct physical processes that relate the inner sources and the surface

P. Ståhle and K.G. Sundin (eds.), IUTAM Symposium on Field Analyses for Determination of Material Parameters - Experimental and Numerical Aspects, 123–137.

data, there holds a certain linearity between them and the relation can be expressed at least formally in terms of the Green function of the semi-infinite body.

An inverse problem of identifying the inner source from the surface data for a semi-infinite body is usually regarded as ill-posed. This is well understood since the effects of the inner source upon the surface responses become smaller as the source is located deeper. The corresponding Green function of the semi-infinite body represents such decaying effects of the source on the data. Therefore, the analysis of Green's function is needed to understand the ill-posedness of the semi-infinite inverse problem and to predict the most reliable solution that can be obtained by using the measured data.

While Green's function is obtained as a solution of a boundary value problem of the semi-infinite body, we can regard Green's function as an operator, a (linear) mapping from a function representing the inner source to a function representing the surface data; see Reed and Simon (1972); see also Kirch (1996) and Hassani (1998). Applying the function analysis theory, therefore, we can compute the *spectral decomposition* of Green's function. This decomposition is similar to the eigen-value decomposition of a square matrix or the singular decomposition of a general matrix. Green's function is thus *discretized* as the sum of distinct functions, even though Green's function is an operator of functions which have infinitely many degree-of-freedom.

In order to solve a semi-infinite body inverse problem, we propose an inverse analysis method that takes advantage of the spectral decomposition of Green's function; see Hori (2000). The key issue is that if Green's function is decomposed, it is straightforward to define a suitable inverse operator which determines the inner source from the surface data. The suitableness means that we can choose the inverse operator which can find a higher mode of the inner source, depending on the accuracy of the measurement. By choosing the inverse operator, we can clearly distinguish modes of the inner source that can be identified from measurements and modes which cannot be identified but will be needed to make, say, the inner source more smoothly distributed.

It should be pointed out that the semi-infinite plane considered is not necessarily homogeneous; the crust is heterogeneous, consisting of several geological layers. For heterogeneous semi-infinite body, we cannot obtain an explicit form of Green's function. We can still apply the spectral decomposition even though Green's function is obtained numerically. Also, it should be emphasized that Green's function can be decomposed before actual measurement is made. The decomposition does not depend on the number or the location of the measuring points.

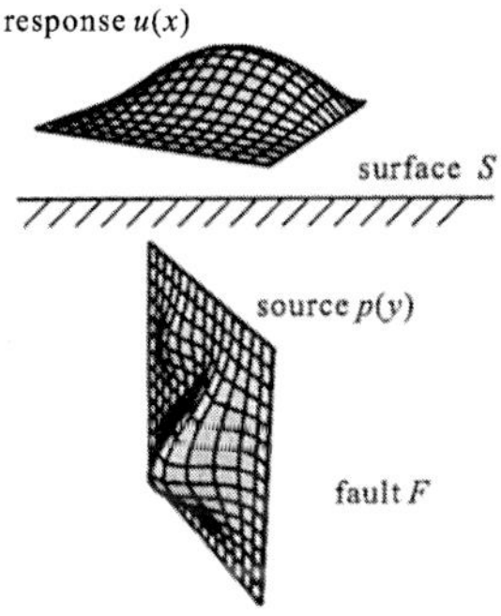

Figure 1 Problem setting.

The content of this paper is as follows: In Section 2, we formulate the inverse analysis method using the spectral decomposition of Green's function. Some emphases are put on the nature of Green's function of the semi-infinite body and the procedure of the inverse method. In Section 3, two-dimensional example problems are solved. The proposed method is compared with simple inverse methods, and it is shown that these methods have errors which may be overlooked.

2. FORMULATION

In this section, we formulate the inverse analysis method using the spectral decomposition of Green's function (Hori, 2000). The target inverse problem is stated as follows:

> Let V be a two- or three-dimensional semi-infinite body. Inner sources p are distributed in a part of the body, F, and surface data u are measured on a part of the body boundary, S. Predict a distribution of p on F when u are measured at some points on S.

See Fig. 1 for a schematic view of the problem setting mentioned above. It is assumed that Green's function of V is given. That is, p and u are related through g as

$$u(\mathbf{x}) = \int_S g(\mathbf{x}, \mathbf{y})p(\mathbf{y}) \, d\mathbf{y}. \tag{1.1}$$

Here, for simplicity, we set u and p as a scalar-valued function and S and F as an arc (for the two-dimensional problem) or an area (for the three-dimensoinal probelm). A case when u or p is a vector-value function or a case when F is a domain within V is treated in the same manner.

2.1 Spectral decompoition and inverse operator

As mentioned, we regard Green's function g as a linear operator between two function spaces. That is, using the same notation F and S to denote a class of functions for the inner source and the surface data, respectively, g maps p in F to u in S. This g must be *compact* since the mapped u becomes a smooth bounded function. According to the function analysis theory, we can assume that g is decomposed as follows:

$$g(\mathbf{x}, \mathbf{y}) = \sum_{\alpha=1}^{\infty} \lambda^\alpha \phi^\alpha(\mathbf{x}) \psi^\alpha(\mathbf{y}). \tag{1.2}$$

Here, $\{\lambda^\alpha, \phi^\alpha, \psi^\alpha\}$ are the eigen-values and eigen-functions that satisfy

1 The eigen-values λ^α vanishes 0 as α increases such that u becomes bounded.

2 The eigen-functions $\{\phi^\alpha\}$ and $\{\psi^\alpha\}$ form an orthonomal base of the function space S and F.

Note that the orthonormality means

$$\int_S \phi^\alpha(\mathbf{x}) \phi^\beta(\mathbf{x}) \, \mathrm{d}\mathbf{x} = \int_F \psi^\alpha(\mathbf{y}) \psi^\beta(\mathbf{y}) \, \mathrm{d}\mathbf{y} = \delta_{\alpha\beta},$$

where $\delta_{\alpha\beta} = 1$ for $\alpha = \beta$ and 0 for $\alpha \neq \beta$. As is seen from Eq. (1.2), g maps ψ^α to $\lambda^\alpha \phi^\alpha$, i.e., $\lambda^\alpha \phi^\alpha = \int_F g \psi^\alpha \, \mathrm{d}\mathbf{y}$. Therefore, denoting $p^\alpha = \int_F \psi^\alpha p \, \mathrm{d}\mathbf{y}$ and $u^\alpha = \int_S \phi^\alpha u \, \mathrm{d}\mathbf{x}$, we have $u^\alpha = \lambda^\alpha p^\alpha$.

In view of Eq. (1.2), we can see the ill-posedness of the semi-infinite body inverse problem. Since λ^α vanishes, we cannot take the infinite summation of $(1/\lambda^\alpha)\psi^\alpha(y)\phi^\alpha(x)$. We have to truncate the summation depending on the accuracy of the measurement, and define the following inverse operator:

$$h(\mathbf{y}, \mathbf{x}) \approx \sum_{\alpha=1}^{K} \frac{1}{\lambda^\alpha} \psi^\alpha(\mathbf{y}) \phi^\alpha(\mathbf{x}), \tag{1.3}$$

where K is chosen such that λ^K/λ^1 does not become smaller than the relative accuracy of the measurement. This h then maps u to p, as

$$p(\mathbf{y}) = \int_S h(\mathbf{y}, \mathbf{x}) u(\mathbf{x}) \, \mathrm{d}\mathbf{x}, \tag{1.4}$$

or $p^\alpha = u^\alpha/\lambda$ for $\alpha = 1, \cdots, K$.

2.2 Computation of spectral decomposition

Computing the eigen-values and eigen-functions, $\{\lambda^\alpha, \phi^\alpha, \psi^\alpha\}$, requires some numerical efforts, in particular, when g is given numerically. This is because g is not self-adjoint as it is a mapping from F to S. To compute $\{\lambda^\alpha, \phi^\alpha, \psi^\alpha\}$, therefore, we first have to define two self-adjoint operators from g, the first for S to S and the second for F to F, as follows:

$$\begin{cases} r(\mathbf{x},\mathbf{x}') = \int_F \; g(\mathbf{x},\mathbf{y})g(\mathbf{x}',\mathbf{y}) \; d\mathbf{y}, \\ \ell(\mathbf{y},\mathbf{y}') = \int_S \; g(\mathbf{x},\mathbf{y})g(\mathbf{x},\mathbf{y}') \; d\mathbf{x}. \end{cases} \tag{1.5}$$

These operators are self-adjoint because they satisfy $r(\mathbf{x},\mathbf{x}') = r(\mathbf{x}',\mathbf{x})$ and $\ell(\mathbf{y},\mathbf{y}') = \ell(\mathbf{y}',\mathbf{y})$. Due to this self-adjointness, the eigen-values and eigen-functions of g are given as solutions of

$$\begin{cases} \lambda^2\phi(\mathbf{x}) = \int_S \; r(\mathbf{x},\mathbf{x}')\phi(\mathbf{x}') \; d\mathbf{x}', \\ \lambda^2\psi(\mathbf{y}) = \int_F \; \ell(\mathbf{y},\mathbf{y}')\psi(\mathbf{y}') \; d\mathbf{y}'. \end{cases} \tag{1.6}$$

The task left is to solve Eq. (1.6) numerically. Even if g is explicitly given in a closed form, it is almost impossible to analytically solve these integral equations. Therefore, we discretize the operators r and ℓ. The discretization of operators is made by the integration. That is, denoting by $\{\xi_i\}$ and $\{\eta_j\}$ a set of M and N functions on F and S, respectively, we can discretize r and ℓ as

$$\begin{cases} r(\mathbf{x},\mathbf{x}') \approx \sum_{i,i'=1}^{M} \; r_{ii'}\xi_i(\mathbf{x})\xi'_{i'}(\mathbf{x}'), \\ \ell(\mathbf{y},\mathbf{y}') \approx \sum_{j,j'=1}^{N} \; \ell_{jj'}\eta_j(\mathbf{y})\eta'_{j'}(\mathbf{y}'), \end{cases} \tag{1.7}$$

where $r_{ii'}$ and $\ell_{jj'}$ are

$$\begin{cases} r_{ii'} = \int_{S\times S} \; r(\mathbf{x},\mathbf{x}')\xi'_i(\mathbf{x})\xi_{i'}(\mathbf{x}') \; d\mathbf{x}d\mathbf{x}', \\ \ell_{jj'} = \int_{F\times F} \; \ell(\mathbf{y},\mathbf{y}')\eta'_j(\mathbf{y})\eta_{j'}(\mathbf{y}')d\mathbf{y}d\mathbf{y}', \end{cases} \tag{1.8}$$

and $\{\xi'_i\}$ and $\{\eta'_j\}$ are functions that are conjugate to $\{\xi_i\}$ and $\{\eta_j\}$, respectively, i.e., $\int_S \; \xi_i(\mathbf{x})\xi'_j(\mathbf{x}) \; d\mathbf{x} = \int_F \; \eta_i(\mathbf{y})\eta'_j(\mathbf{y}) \; d\mathbf{y} = \delta_{ij}$. Since ϕ and ψ are discretized as $\phi_i = \int_S \; \xi_i\phi \; d\mathbf{x}$ and $\psi_j = \int_F \; \eta_j\psi \; d\mathbf{y}$, the eigen-value problems, Eq. (1.6), is now discretized as $\lambda^2[\phi_i] = [r_{ii'}][\phi_{i'}]$ and $\lambda^2[\psi_j] = [\ell_{jj'}][\psi_{j'}]$. It should be noted that the matrices $[r_{ii'}]$ and $[\ell_{jj'}]$ are not symmetric, while the operators r and ℓ are self-adjoint as defined by Eq. (1.5). However, the matrices are the proper discretization of the operators; for instance, the following correspondence holds for r and $[r_{ii'}]$:

$$a(\mathbf{x}) = \int_S \; r(\mathbf{x},\mathbf{x}')b(\mathbf{x}') \; d\mathbf{x}' \quad \Longrightarrow \quad [a_i] \approx [r_{ii'}][b_{i'}], \tag{1.9}$$

for $a_i = \int_S \; a\xi_i \; d\mathbf{x}$ and $b_i = \int_S \; b\xi_i \; d\mathbf{x}$.

2.3. Inverse analysis method using spectral decomposition

Now, we propose an inverse analysis method using the spectral decomposition of Green's function. The key issue is that the surface data u are not given as a continuous function but values measured at, say, M points $\{\mathbf{x}_i\}$, i.e., $\{U_i\} = \{u(\mathbf{x}_i)\}$. We need to reconstruct a function u from the data $\{U_i\}$.

The task of reconstructing u from $\{U_i\}$, however, is not so difficult if Green's function is decomposed and the form of u is assumed as

$$u^*(\mathbf{x}) = \sum_{\alpha=1}^{K^*} u^{*\alpha}\phi^\alpha(\mathbf{x}), \tag{1.10}$$

where K^* is the number of eigen-functions which can be measured; for instance, we can set $K^* = K$ which determines the inverse operator h as shown by Eq. (1.3). We can find the following two ways to determine the coefficients $u^{*\alpha}$ from $\{U_i\}$:

1 Interpolate or extrapolate $\{U_i\}$ using suitable spline functions f_i for the measured point $\mathbf{x}_i$. The coefficient is then computed as $u^{*\alpha} = \int_S \phi^\alpha \left(\sum_{i=1}^M U_i f_i\right) \,\mathrm{d}\mathbf{x}$.

2 Define the error u as $E(\{u^{*\alpha}\}) = \sum_{i=1}^M (U_i - \sum_{\alpha=1}^{K^*} u^{*\alpha}\phi^\alpha(\mathbf{x}_i))^2$, where u^* is given by Eq. (1.10). The coefficients $\{u^{*\alpha}\}$ are obtained by minimizing E.

Note that we can compute ϕ^α using the discretized vector $[\phi_i^\alpha]$, i.e., $\phi^\alpha = \sum_{i=1}^M \phi_i^\alpha \xi_i$.

Once $\{u^{*\alpha}\}$ are determined, the inner source can be explicitly expressed in terms of the eigen-values and eigen-functions, $\{\lambda^\alpha\}$ and $\{\psi^\alpha\}$, as

$$p(\mathbf{y}) = \sum_{\alpha=1}^{K^*} \frac{u^{*\alpha}}{\lambda^\alpha}\psi^\alpha(\mathbf{y}). \tag{1.11}$$

This p is obtained by operating the inverse operator h to the reconstructed u^*; see Eq. (1.4).

As summarized in Fig. 2, the proposed inverse analysis method has the following three steps: 1) compute the spectral decomposition of g suitably discretizing the operators r and ℓ; 2) reconstruct the surface data function u from the measured surface data $\{U_i\}$; and 3) obtain the inner source function p using Eq. (1.11).

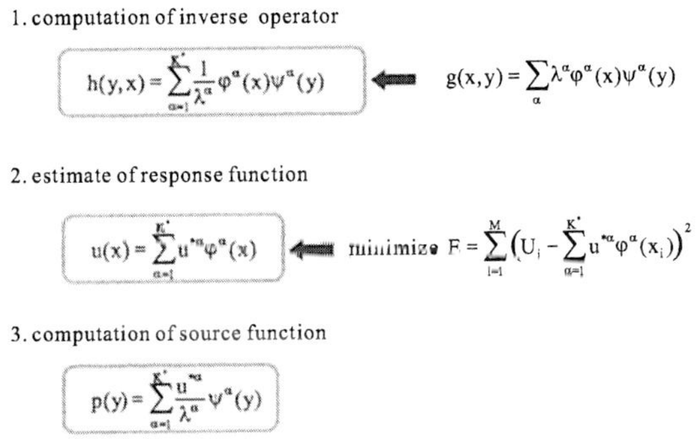

Figure 2 Summary of inverse analysis method using spectral decomposition of Green's function.

2.4 Approximation of computing spectral decomposition

The numerical computation required to make the spectral decomposition of Green's function, or to solve Eq. (1.6), is not trivial, since it requires the determination and the discretization of r and ℓ, Eqs. (1.5) and (1.8), which need the integration over S and F. Such computation could be reduced if we approximately solve Eq. (1.6).

For instance, we can apply the Fourier series expansion to g when F and S are a line segment (for the two-dimensional problem) or a rectangular domain (for the three-dimensional problem), even though a large number of terms will be required to accurately compute g. The coefficient of the expansion is given as $g_{ij}^{FE} = \int_{S\times F} g(\mathbf{x},\mathbf{y})\exp(\imath(\mathbf{k}_i\cdot\mathbf{x}+\mathbf{l}_j\cdot\mathbf{y}))\,\mathrm{d}\mathbf{x}\mathrm{d}\mathbf{y}$, where $\mathbf{k}_i$ and $\mathbf{l}_j$ are the Fourier series variables for $\mathbf{x}$ and $\mathbf{y}$ and $\cdot$ stands for the inner product. Due to the orthonormality of $\exp(\imath\mathbf{k}_i\cdot\mathbf{x})$ and $\exp(\imath\mathbf{l}_i\cdot\mathbf{x})$, the matarices corresponding to the discretized operators r and ℓ are$[g_{i'j}^{FE}][g_{ij}^{FE}]^T$ and $[g_{ij}^{FE}]^T[g_{ij'}^{FE}]$.

In a similar manner, we can discretize g using $\{\xi_i\}$ and $\{\eta_j\}$ instead of $\exp(\imath\mathbf{k}_i\cdot\mathbf{x})$ and $\exp(\imath\mathbf{l}_i\cdot\mathbf{x})$, i.e.,

$$g_{ij} = \int_{S\times F} g(\mathbf{x},\mathbf{y})\xi_i(\mathbf{x})\eta_j(\mathbf{y})\,\mathrm{d}\mathbf{x}\mathrm{d}\mathbf{y}. \tag{1.12}$$

The discretization of the operators r and ℓ becomes of more complicated form than the Fourier series expansion, since $\{\xi_i\}$ and $\{\eta_j\}$ are not orthonomal bases. Indeed, denoting by $I_{ii'}^{\xi\xi} = \int_S \xi_i\xi_{i'}\,\mathrm{d}\mathbf{x}$ and $I_{jj'}^{\eta\eta} = \int_F \eta_j\eta_{j'}\,\mathrm{d}\mathbf{y}$, we can approximately compute $[r_{ii'}]$ and $[\ell_{jj'}]$ given by Eq. (1.8), as follows:

$$\begin{cases} [r_{ii'}] \approx \left[I^{\xi\xi}_{ik}\right]^{-1} [g_{kj}] \left[[I^{\eta\eta}_{jj'}\right]^{-1} [g_{i'j'}]^T, \\ [\ell_{jj'}] \approx \left[I^{\eta\eta}_{jk}\right]^{-1} [g_{ik}]^T \left[I^{\xi\xi}_{ii'}\right]^{-1} [g_{i'j'}]. \end{cases} \tag{1.13}$$

The complicated form of $[r_{ii'}]$ and $[\ell_{jj'}]$ given by Eq. (1.13) shows errors which may be overlooked if Green's function is discretized without applying the integration. For instance, the simplest discretization of g is point-wise, i.e., $G_{ij} = G(\mathbf{x}_i, \mathbf{y}_j)$ where $\{\mathbf{x}_i\}$ are the measuring points and $\{\mathbf{y}_j\}$ are points suitably chosen on F. Denoting the characteristic functions of S and F by δ^S and δ^F, respectively, we can rewrite G_{ij} in the same form as Eq. (1.12), as

$$G_{ij} = \int_{S\times F} g(\mathbf{x}, \mathbf{y})\delta^S(\mathbf{x}-\mathbf{x}_i)\delta^F(\mathbf{y}-\mathbf{y}_j)\ \mathrm{d}\mathbf{x}\mathrm{d}\mathbf{y}. \tag{1.14}$$

The simplest point-discretization of Eq. (1.1) leads to $[U_i] = [G_{ij}][P_j]$.

As an alternative of the point-wise discretization, we may use spline functions for the inner source function even though the surface data function is discretized point-wisely. Such discretization, however, may lead to some error in solving the inverse problem, since this discretization of Green's function is expressed in the form of

$$G'_{ij} = \int_{S\times F} g(\mathbf{x}, \mathbf{y})\delta^S(\mathbf{x}-\mathbf{x}_i)\eta_j(\mathbf{y})\ \mathrm{d}\mathbf{x}\mathrm{d}\mathbf{y}, \tag{1.15}$$

where $\{\eta_j\}$ are regarded as the spline function.

Using either Eq. (1.14) or (1.15), respectively, we can obtain a matrix equation, $[U_i] = [G_{ij}][P_j]$ or $[U_i] = [G'_{ij}][P'_j]$, where $[U_i]$ is the measured surface data and $[P_j]$ or $[P'_j]$ is the discretized inner source function. A solution can be found for this matrix equation, say, by computing the singular value decomposition of the matrix $[G_{ij}]$ or $[G'_{ij}]$. For instance, the decomposition of $[G_{ij}]$ uses two square symmetric matrices,

$$\begin{cases} [R_{ii'}] = [G_{ij}]\,[G_{i'j}]^T, \\ [L_{jj'}] = [G_{ij}]^T\,[G_{ij'}]. \end{cases} \tag{1.16}$$

It should be emphasized that the form of Eq. (1.16) is different from the form of Eq. (1.13) which is derived from the approximate but proper discretization of the operators r and ℓ; see Eq. (1.9). Thus, $[P_i]$ obtained may not be a proper discretization of the inner source function, even though it produces a negligibly small error of $||[U_i] - [G_{ij}][P_j]||$ with $||$ being the norm of the vector. If the singular value decomposition of $[G'_{ij}]$ is used, $[P'_j]$ may have similar error. This will be studied in the next section.

3. EXAMPLE PROBLEM

We study an example problem in the two-dimensional setting, where the semi-infinite body is $x_2 > 0$ and S and F are $S = \{(x, 0) \mid s_{mn} < x < s_{mx}\}$ and $F = \{(0, y) \mid f_{mn} < y < f_{mx}\}$. For simplicity, we consider a potential problem; the inner sources are double-couples and the surface data are potentials. Assuming the free flux boundary conditions on ∂V, we have $g(x, y) = -(1/\pi)x/(x^2 + y^2)$. Decomposing S and F into M and N segments with equal length dx and dy, we set $\{\xi_i\}$ and $\{\eta_j\}$ as piece-wise linear functions on these segments. The nodes of the segments are denoted by x_i and y_j $(i = 0, 1, \cdots, M,\ j = 0, 1, \cdots, N)$.

First, we verify the validity of numerically computing the spectral decomposition of g. For $(s_{mn}, s_{mx}) = (1, 3)$ and $(f_{mn}, f_{mx}) = (1, 3)$, we examine the convergence of the eigen-values and eigen-functions by increasing the number of the segments. Five cases of $N = M = 5$, 10, 20, 40 and 80, are tested. The eigen-values and eigen-functions are presented in Table 1 and Fig. 3, respectively. For comparison, we present those obtained by using the singular value decomposition of $[R_{ii'}]$ and $[L_{jj'}]$. As is seen, the eigen-values given by $[r_{ii'}]$ and $[\ell_{jj'}]$ coincide with each other and converge well. Also, the eigen-functions given by $[r_{ii'}]$ and $[\ell_{jj'}]$ converge fast. It should be noted that unlike the eigen-values, $[R_{ii'}]$ and $[L_{jj'}]$ produce eigen-functions similar to those of $[r_{ii'}]$ and $[\ell_{jj'}]$.

Table 1 Convergence of eigen-values.

a) absolute value

N	r	l	R
5	0.16675470	0.16675518	0.42874550
10	0.16675569	0.16675569	0.84400444
20	0.16675572	0.16675572	1.67703995
40	0.16675572	0.16675572	3.34425627

b) λ^2

N	r	l	R
5	0.01350124	0.01349571	0.05042901
10	0.01350223	0.01350203	0.08300752
20	0.01350227	0.01350226	0.14982390
40	0.01350227	0.01350227	0.28449842

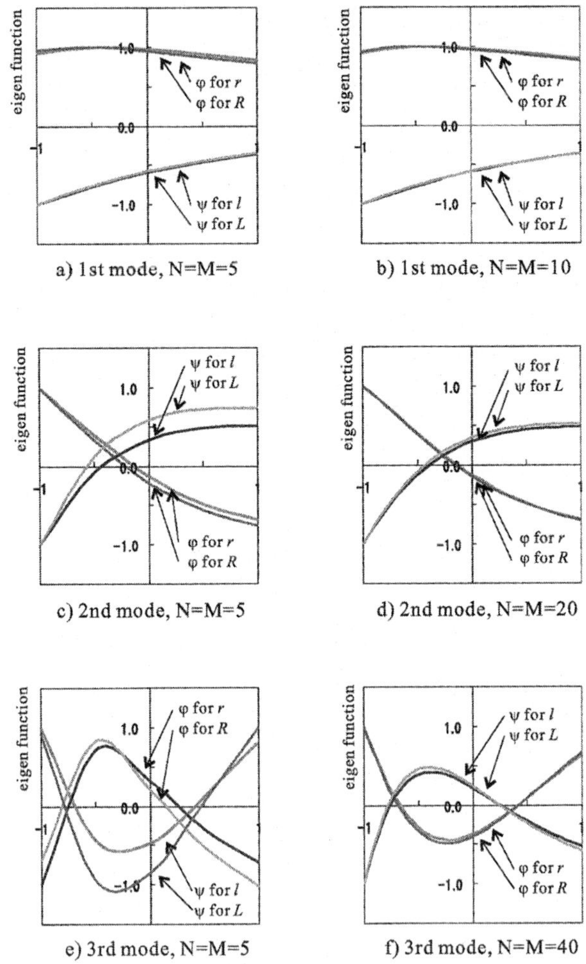

Figure 3 Convergence of eigen-functions.

As expected, the eigen-values decrease exponentially as shown in Table 1. Thus, even a small error may cause a large error in computing the eigen-values. Figure 4 presents a black-and-white images of the matrices $[r_{ii'}]$ and $[R_{ii'}]$ for $N = M = 5$ to demonstrate this; components are divided by their maximum value for the comparison. The difference between $[r_{ii'}]$ and $[R_{ii'}]$ is not significant.

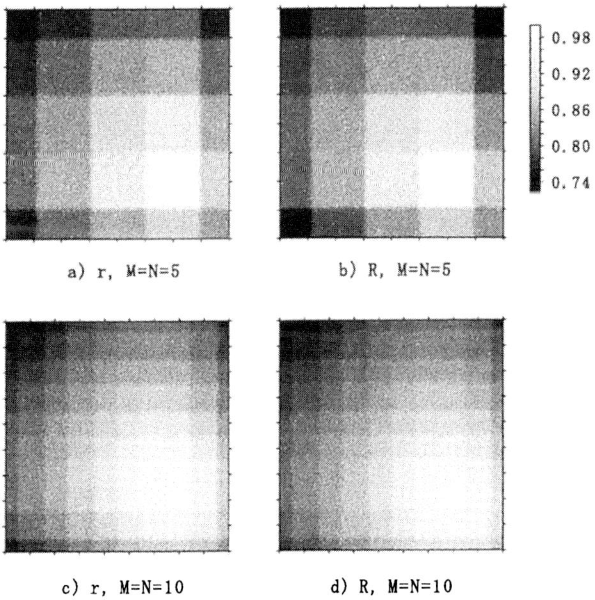

Figure 4 Difference in matrices $[r_{ii'}]$ and $[R_{ii'}]$.

Table 2 Problem setting for comparison with simplest method.

	1	2	3	4	5
set 1	1.00	1.25	1.50	2.00	3.00
set 2	1.00	1.75	2.00	2.25	3.00

Now, we compare the proposed inverse analysis method with the simplest inverse method using point-wise discretization. The domain is $(s_{mn}, s_{mx}) = (1, 3)$ and $(f_{mn}, f_{mx}) = (1, 3)$, and two sets of five measuring points are used; see Table 2. The new method uses ten segments for both S and F, and the simplest inversion method uses eleven nodes.

We first compare the eigen-values. Table 3 shows first three eigen-values $(\lambda^{1,2,3})$ and the relative value of $\lambda^{\alpha}/\lambda^{1}$. The simplest method produces the eigen-values which depend on the measuring points. However, the relative values of the eigen-values are more or less the same as those obtained by the new method. These show the error inherent

Table 3 Comparison of eigen-values.

a) absolute values

	new method	set 1	set 2
λ^1	0.166755720	1.182634348	1.170764707
λ^2	0.013502267	0.114516942	0.109383043
λ^3	0.000918963	0.009508547	0.009847124

b) relative values

	new method	set 1	set 2
λ^2/λ^1	8.097034	9.683208	9.342872
λ^3/λ^1	0.551083	0.804014	0.841085

to the simplest method, if it is regarded as an approximate method of computing the spectral decomposition of Green's function.

We plot the three eigen-functions in Fig. 5; the eigen-functions of the simplest method are obtained by smoothly connecting the components of the eigen-vector of $[R_{ii'}]$ or $[L_{jj'}]$. The eigen-functions of S given by the simplest method change depending on the location of the measurement, like the eigen-values. The first eigen-function of F, however, is accurately computed as it is similar to that obtained by the new method. The second and third eigen-functions are less accurately computed.

Since the eigen-values of g decreases exponentially, a small error in computing the eigen-values and eigen-functions results in a large error in inverting the inner source from the measured data. To clarify such an error quantitatively, we simulate the inversion of predicting the source function, assuming that surface data are measured at five points as shown in Table 2. The present method is compared with the simplest inversion method that uses the point-wise discretization and the inversion method that uses the smooth source function; see Eq. (1.16), for the discretization of the operators r and ℓ.

We study a case when the source function is given by the first eigen-function, i.e., $p(y) = \psi^1(y)$. The surface data function becomes $u(x) = \lambda^1\phi^1(x)$, and five data are obtained. According to the present method, the inversion is carried out as follows: 1) set $K^* = 3$ in view of $\lambda^3/\lambda^1 \approx 1\%$; 2) obtain $u^{*\alpha} = \lambda^1$ for $\alpha = 1$ or 0 for $\alpha \neq 1$ minimizing the error E; and 3) obtain p substituting $\{u^{*\alpha}\}$ into Eq. (1.11). In this case, we can accurately compute $\{u^{*\alpha}\}$ since $\{\phi^\alpha\}$ are orthonormal. Figure 6 plots the inner source obtained. Also, the predictions of the other two method

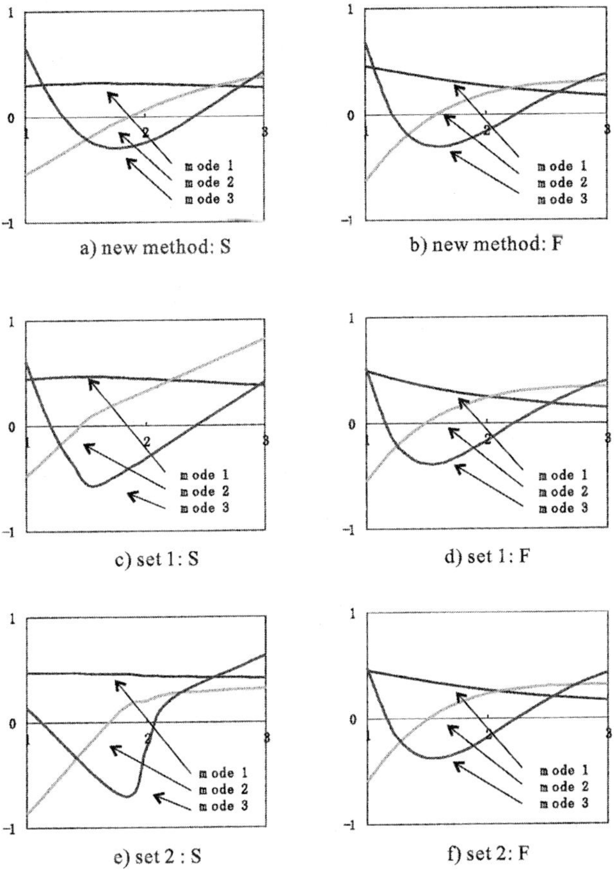

Figure 5 Comparison of eigen-functions.

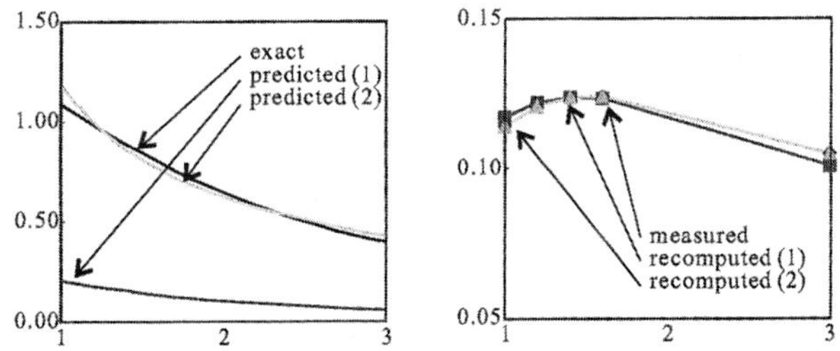

Figure 6 Predicted source function and recomputed response function.

are presented. It is seen that the simplest method fails in inverting p. However, if we recompute u using the predicted p, the recomputed u is almost the same as the accurate one; see Fig. 6. This means that $\|[U_i] - [G_{ij}][P_j]\|$ is almost zero since the simplest inversion method obtains a solution of the matrix equation, $[U_i] - [G_{ij}][P_j]$ although it cannot solve $u = \int_S gp \, \mathrm{d}\mathbf{x}$.

It is shown in Fig. 6 that the inversion method using the smooth source function works well. However, there are cases when the method fails due to less accurate computation of the spectral decomposition; see Hori (2000). This could be inevitable if we see the form of $[G'_{ij}]$ given by Eq. (1.15) where δ^F is used; $[R_{ii'}]$ or $[L_{jj'}]$ given by this $[G_{ij}]$ is not proper discretization of the operators r and ℓ in a sense of Eq. (1.9).

4. CONCLUDING REMARKS

For a semi-infinite body inverse problem, we propose a new analysis method which uses the spectral decomposition of Green's function. While the present method requires some amount of numerical computation, it can determine a suitable inverse operator and predict an inner source function from a set of surface data. Example problems show the validity of the present method. Also, it is shown that simple inverse analysis methods have errors which may be overlooked. These errors are due to the improper discretization of the operator.

Acknowledgments

This research is supported partially by Grant-in-Aid for Scientific Research, the Ministry of Education, Science, Sports and Culture and partially by Japan Science and Technology Corporation.

References

Dimri, V. (1992). *Deconvolution and Inverse Theory: Application to Geophysical Problems*, Elsevier, Amsterdam.

Fowler, C.M.R. (1990). *The Solid Earth: An Introduction to Global Geophysics*, Cambridge University Press, Cambridge.

Hassani, S. (1998). *Mathematical Physics: A Modern Introduction to Its Foundations*, Springer, New York.

Hornby, P., Boshetti, F. and Horowiz, F.G. 1999. Analysis of potential field data in the wavelet domain, *Geophys. J. Int.*, **137**, 175-196.

Hori, M. (2000). Inverse analysis method using spectral decomposition of Green's function, *Geophys. J. Int.* (to appear)..

Kirch, A. (1996). *An Introduction to the Mathematical Theory of Inverse Problems*, Springer, New York.

Menke, W. (1989). *Geophysical Data Analysis*, Academic Press, New York.

Ranbothen, P.S. and Pratt, R.G. (1997). Improved inversion through use of the null space, *Geophys.*, **62**, 869-883.

Reed, M. and Simon, B. (1972). *Methods of Modern Mathematical Physics, Vol. I: Function Analysis*, Academic Press, New York.

Sen, M.K. and Stoffa, P.L. (1995). *Global Optimization Methods in Geophysical Inversion*, Elsevier, Amsterdam.

Tarantola, A. (1987). *Inverse Problem Theory*, Elsevier, Amsterdam.

Yao, Z.S. and Roberts, R.G. (1999). A practical regularization for seismic tomography, *Geophys. J. Int.*, **138**, 293-299.

DETERMINATION OF BALLAST PARAMETERS OF A RAILWAY TRACK USING AN OPTIMIZATION TECHNIQUE

V.L. Markine, C. Esveld
Section of Road and Railway Engineering
Faculty of Civil Engineering, Delft University of Technology
Stevinweg 1, NL-2628, CN Delft, The Netherlands

Abstract: The paper presents a procedure for analysis of lateral resistance of a railway track. The lateral load/displacements diagrams have been obtained by shifting a real track using a tamping machine, which is usually used to correct the vertical geometry of a track. The measurements are simulated using a 3-D FE model of a track. Based on experimental data and results of numerical simulation, the model parameters related to ballast lateral resistance are determined using an optimisation technique. The ballast parameter identification procedure has been verified using measurement data with and without extra vertical load applied to a track. Finally, a track model with the obtained ballast parameters has been applied to determine train speed limits for a renewed track basing on the safety criteria given in UIC Leaflet 720. The results are presented and discussed.

1. INTRODUCTION

The lateral strength of a railway track is to a large extent defined by ballast lateral resistance. For a newly laid track or a track after full maintenance, ballast particles are not good enough consolidated and therefore lateral resistance of such a track is poor. This can be observed from *Figure 1* where a typical behavior of well-consolidated and tamped ballast during a single sleeper test is shown. In order to prevent track buckling, lateral forces occurring during operations on newly laid tracks or tracks after maintenance should be restricted. Usually, this is achieved by temporary restriction of train speeds on renewed sections of a track. As a number of train passes grows, ballast becomes more consolidated and its lateral resistance increases. Accordingly, the restriction of the speed limits can be released.

P. Ståhle and K.G. Sundin (eds.), IUTAM Symposium on Field Analyses for Determination of Material Parameters - Experimental and Numerical Aspects, 139–149.

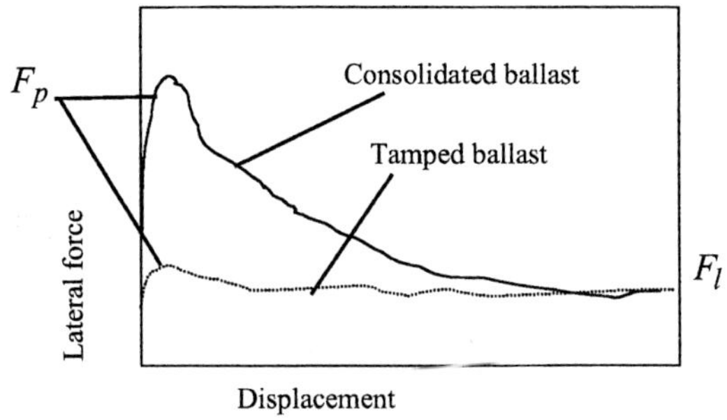

Figure 1. Lateral behaviour of ballast in a single sleeper test (not to scale)

Obviously, an introduction of temporary speed restrictions deteriorates operational efficiency of a railway track. To apply proper speed limits, knowledge about lateral resistance of a track and its interpretation for determination of safe speed limits are required. That is why, a technique for identification of ballast lateral resistance parameters basing on measured lateral resistance of a real track has been developed at Delft University of Technology, so that the UIC safety criteria could be applied for determination of speed limits [3]. The procedure is described below.

2. MEASUREMENT OF LATERAL RESISTANCE OF TRACK

A technique for measurement of lateral resistance of a track employed by Nederlandse Spoorwegen (Netherlands Railways) is used here. The technique makes use of a tamping machine, which, instead of lifting, shifts a track frame in the lateral direction. Resulting lateral displacements of a track frame are measured and combined in a force-displacement diagram as shown in *Figure 2*. An applied lateral force is slowly increased resulting in movement of a track frame (path O-A-B in *Figure 2*). As the lateral displacements have achieved a prescribed maximum value (point B in *Figure 2*) shifting grips are released so that a track frame is moving back (path B-C-D in *Figure 2*). It should be noted that the maximum displacements of a track during such measurements are relatively small (max. 4mm) and residual displacements of the track are not too large so that the original track geometry can easily be restored. The elastic limits of a track can then be obtained by considering the residual displacements of a track frame (distance O-D in *Figure 2*).

The maximum elastic displacements $\overline{W}_p^{ex}$ of the track are estimated by the displacements of a track Δ after the lateral load has been released (path B-D in *Figure 2*), i.e.

$$\overline{W}_p^{ex} = \Delta . \tag{1}$$

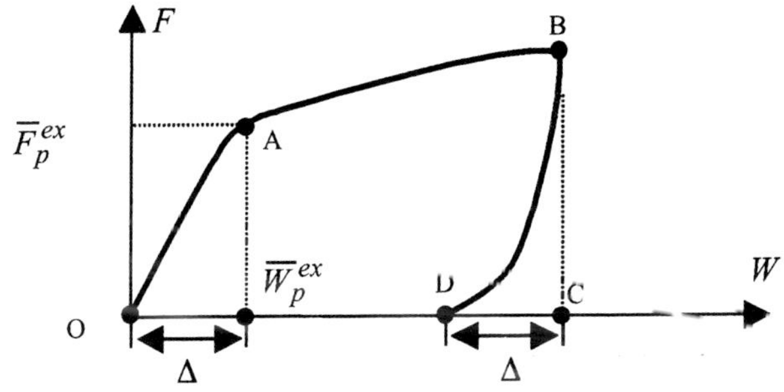

Figure 2. Lateral force-displacement diagram of track measured using tamping machine (not to scale)

The maximum elastic limit $\overline{F}_p^{ex}$ is defined as the load corresponding to the displacement $\overline{W}_p^{ex}$ (*Figure 2*).

3. NUMERICAL MODEL OF TRACK

The 3-D Finite Element (FE) model used here for simulation of track behavior is shown in *Figure 3*. The model can be used for analysis of both straight and curved tracks [3,7] with misalignments (*Figure 3*a). Rails are modeled by 3-D elastic beam

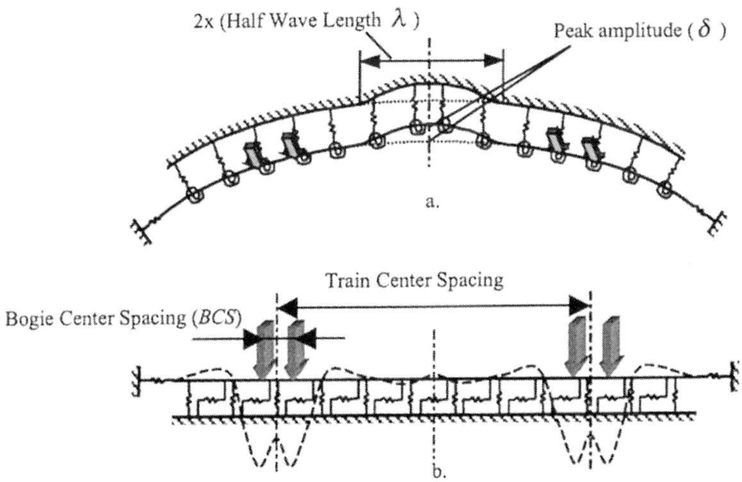

Figure 3. FE model of track: a. Top view; b. Side view

elements. The vertical behavior of ballast is described by linear spring elements (*Figure 3*b), which provide a support to rails also known as the Winkler foundation. The longitudinal behavior of a track is assumed to be linear and therefore is also modeled using linear spring elements (*Figure 3*b). Combined torsional stiffness of a track defined by fastening system, track gauge and sleeper spacing is approximated by linear torsional springs (*Figure 3*a). The choice between linear and non-linear spring elements was based on the experimental data of various tracks available from the literature [6,7].

The lateral behavior of ballast is non-linear (*Figure 1*) to model which elasto-plastic spring elements have been used. The elements can describe bi-linear or non-linear with softening behavior as shown in

Figure 4. The linear and non-linear (model with softening) branches of the element behavior are respectively described by the functions

$$s = K_{el}\, w \qquad \text{and} \qquad s = F_l + (F_p - F_l)2^{-w/W_l}\ , \tag{2}$$

where W_p and F_p are the elastic limits of ballast for unloaded (reference) case, and $K_{el} = F_p / W_p$ is the elastic stiffness. The rate of softening is defined by the parameters $W_l > W_p$ and $F_l > F_p$ (See *Figure 4* a). These are chosen so that the load s corresponding to the displacement $w = W_l$ is equal to the average value between F_p and F_l. From (2) it can be seen that if $F_l = F_p$ the model describes bi-linear behavior of ballast as shown in *Figure 4*b.

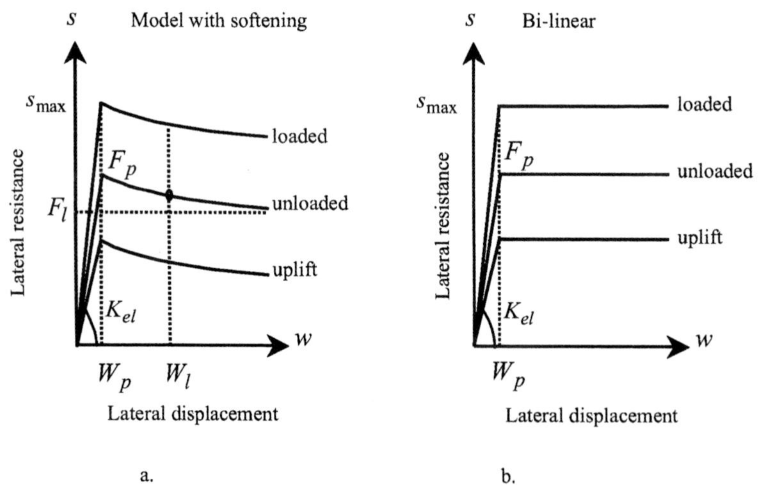

Figure 4. Lateral ballast behaviour (model): a. with softening; b. bi-linear

From measurements it is known that ballast behavior depends on vertical loads. The model accounts for the effect of vertical forces by using the Mohr-Coulomb criterion. The yielding of ballast now occurs at the force s_{max} (loaded and uplift plots on

Figure 4) which can be calculated as:

$$s_{max} = F_p - s_v \tan\varphi \tag{3}$$

where s_v is the vertical load per spring element and $\tan\varphi$ is the friction coefficient between the sleeper and ballast.

Thus, if no vertical load is applied, ($s_v = 0$), the yielding of ballast occurs at the lateral force $s_{max} = F_p$. The stiffness of ballast K_{el} also depends on the vertical load and can be obtained from (2) and (3) according to the formula

$$K_{el} = \frac{F_p - s_v \tan\varphi}{W_p}. \tag{4}$$

To derive an expression for the softening branch in case of vertical loading one can notice that the elastic (load) limit of ballast is changing by the factor $b = s_{max} / F_p$. Assuming the linear dependence between lateral resistance and vertical load, the limit parameter F_l' in case of a vertical load is also increasing by the same factor b, i.e. $F_l' = F_l \frac{s_{max}}{F_p}$. Introducing the parameter $s_{lim} = F_l / F_p$ and using (2) the softening branch in general case can be described by the formula

$$s = s_{max}\left[s_{lim} + (1 - s_{lim})2^{-w/W_l}\right] \tag{5}$$

Thus, a bi-linear model is characterized by the three parameters F_p, W_p and $\tan\varphi$ whereas a model with softening is defined by five parameters, namely F_p, F_l, W_p, W_l and $\tan\varphi$. A model with softening can accurately describe ballast behavior but its use requires a non-linear static response analysis, which can be time consuming. On the other hand, response quantities of a track can be obtained very fast if a bi-linear model of ballast is used, since a response analysis in this case is linear.

The lateral behavior of a railway track can be simulated using the above described numerical model providing that ballast parameters are known. On the other hand, if the lateral behavior of a whole track is known one can try to solve an inverse problem to determine ballast parameters. A procedure for determination of ballast parameters is presented below.

4. BALLAST PARAMETER IDENTIFICATION

Since elastic limits of ballast F_p and W_p have no affect on plastic behavior of a track and to speed up an optimization process, they can be determined separately from the other parameters. Moreover, because of a relatively small shift of a track during measurements and therefore a lack of experimental data about plastic deformations of a track the parameters F_l and W_l characterizing plastic behavior of ballast could not be obtained. That is why determination of these parameters is not discussed here although if experimental data on plastic behavior of a track is available the same technique (as for elastic parameters) can be applied to determine the parameters F_l and W_l. Thus, determination of the ballast parameters F_p and W_p is based on the elastic parameters of a real track $\overline{W}_p^{ex}$ and $\overline{F}_p^{ex}$ obtained from measurements (*Figure 2*). Identification of the ballast parameters is then equivalent to the following minimization problem:

Minimize

$$G(\boldsymbol{a}) \equiv \left| \ddot{\mathbf{O}}(\boldsymbol{X}) - \tilde{\ddot{\mathbf{O}}}(\boldsymbol{X}, \boldsymbol{a}) \right| \rightarrow \min \tag{6}$$

with

$$\boldsymbol{a} = [F_p \; W_p]^T, \boldsymbol{X} = [\overline{F}_p^{ex}], \; \ddot{\mathbf{O}}(\boldsymbol{X}) = [\overline{W}_p^{ex}], \; \tilde{\ddot{\mathbf{O}}}(\boldsymbol{X}) = [\overline{W}_p]. \tag{7}$$

Here, the vectors $\ddot{\mathbf{O}}$ and $\tilde{\ddot{\mathbf{O}}}$ related to responses of a track and numerical model consist of displacements respectively of a track $\overline{W}_p^{ex}$ and numerical model $\overline{W}_p$ at the application point of the lateral load. If displacements of other points of a track are available they can be taken into account as well. When non-linear behavior of a track is analyzed, more loading steps (components of the vector $\boldsymbol{X}$) should be used.

Since numerical responses of a model are obtained performing a static linear analysis the optimization problem (6)-(7) can be solved relatively fast by using a conventional method of non-linear mathematical programming (NMP). Here, a method for minimization of L1 norm of a non-linear function has been used [2].

If no extra vertical load is applied, the solution of the minimization problem (6)-(7) $\boldsymbol{a}^* = [F_p^* \, W_p^*]^T$ can be considered as elastic limits of ballast, i.e. $F_p = F_p^*$ and $W_p = W_p^*$. If measurements have been performed with an extra vertical load F_v the obtained parameter F_p^* is considered as the maximum yielding force $s_{\max} = F_p^*$ defined in (3). The elastic force parameter of ballast F_p (for unloaded case) can then be calculated from (3) as

$$F_p = F_p^* + s_v \tan\varphi \tag{8}$$

where s_v is the vertical load per spring element. The corresponding elastic deformation limit W_p is evaluated as follows:

$$W_p = \frac{F_p}{K_{el}}, \qquad \text{where } K_{el} = \frac{F_p^*}{W_p^*}. \tag{9}$$

It should be noted that the rail temperature during measurements can also be taken into account in ballast parameter identification.

As the parameters of ballast have been determined the numerical model can be used for analysis of track behavior under various loading. In the following section a procedure for determination of train speed limits on renewed tracks is described.

5. DETERMINATION OF SPEED LIMITS

Determination of speed limits is based on the safety criteria given in UIC Leaflet-720 [1]. According to these criteria, the maximum allowable temperature of the rails T_{allow} is defined by two critical buckling temperatures $T_{b,\min}$ and $T_{b,\max}$ (*Figure 5*) as follows

- if $\Delta T > 20°C$, $T_{allow} = T_{b,\min} + 0.25\Delta T$,
- if $5°C < \Delta T < 20°C$, $T_{allow} = T_{b,\min}$,
- if $\Delta T \le 0°C$, $T_{allow} = T_{b,\min} - 5°C$,
- where $\Delta T = T_{b,\max} - T_{b,\min}$.
-

Critical buckling temperatures depend on axle loads and lateral loads due to curvature of track and rail misalignments. The lateral loads can be approximated by the following expression

$$F_{lat} = \frac{\alpha M V^2}{R}, \tag{10}$$

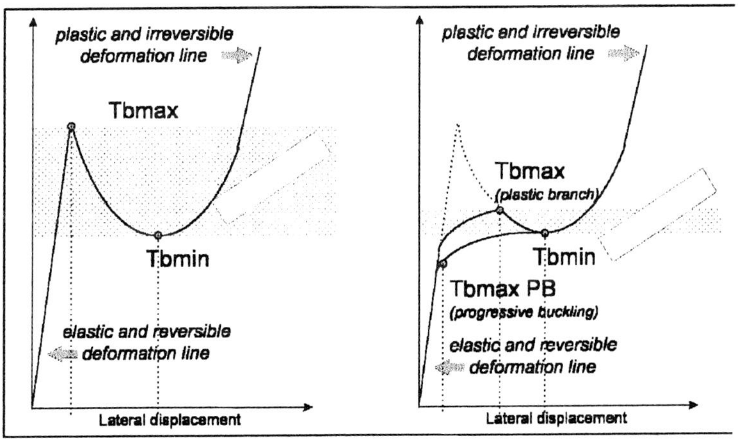

Figure 5. Critical buckling temperatures of rails

where V is the train velocity; $R = \dfrac{2\pi^2\delta}{(2\lambda)^2}$ is the equivalent radius of misalignments (*Figure 3*); α is a coefficient characterizing wheel quality, viz. $\alpha = 1.0$ for a passenger train and $\alpha = 1.5$ for a freight train. The track model (*Figure 3*) has been used to calculate the maximum allowable temperature T_{allow} according to the safety criteria given above. Maximum allowable temperatures $(T_{allow})_i, i = 1,...,N$ calculated for N typical velocities of the train are to be compared with a combined equivalent temperature T_{equiv}, which depends on air temperature, forces due to braking/accelerating of a train etc [1]. The particular velocity of a train V_i is considered to be safe if the difference $T_{allow}\big|_{V_i} - T_{equiv}$ is positive. The maximum train velocity V^* for which the difference $T_{allow}\big|_{V^*} - T_{equiv}$ is still positive can be taken as a speed limit.

6. NUMERICAL EXAMPLES

6.1 Verification of parameter identification procedure

To verify the parameter identification procedure measurement data of a track with wooden sleepers have been used [4]. The data has been obtained employing tamping machines. A friction coefficient for wooden sleepers $\varphi = 1.1$ have been used.

Table 1. Results of identification problem during verification

Input			Output		
$\overline{F}_p^{ex}[N]$	$\overline{W}_p^{ex}[m]$	$T_{\exp}$	$F_p[N/m]$	$W_p[m]$	$K_{el}[N/m/m]$
64800	0.0018	0	44680	0.0018	6928333

Since there was no extra vertical load applied to a track during the measurements ($F_v = s_v = 0N$), the values of the parameters F_p^* and W_p^* obtained from identification have been taken as ballast properties (see Eq. (8)), viz. $F_p = F_p^* = 44680\ N/m$, $W_p = W_p^* = 0.0018\ m$ (*Table 1*). The lateral behavior of ballast (sleeper) and track using the obtained ballast parameters with no extra vertical load is shown in *Figure 6*.

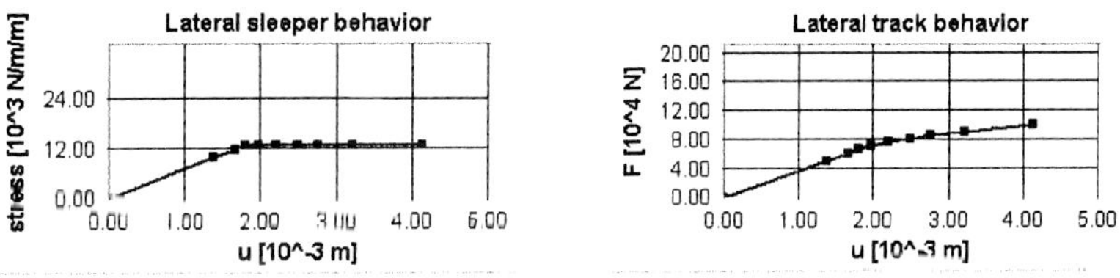

Figure 6. Simulation of measurement with no vertical load

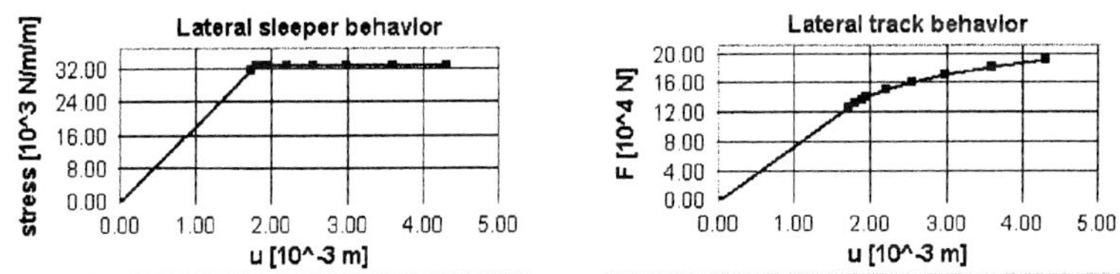

Figure 7. Simulation of measurement with extra vertical load of 100kN

The same model has been used to simulate the measurements performed on the same track but with an extra vertical load $F_v = 100kN$ (applied at the application point of a lateral load) [5]. The results of the simulation are shown in *Figure 7*. Since the simulated maximum elastic limit of a track $F_p = 152\ kN$ (*Figure 7*) is very close to the measured one [5] it was concluded that the identification procedure was correct.

6.2 Determination of speed limits

Now the identification procedure has been applied to determine speed limits on a track after slipper renewing. The lateral resistance of a track has been measured directly after maintenance as well as 3 and 12 days later [5]. Track and ballast elastic limits obtained in [3] using the parameter identification procedure are collected in *Table 2*. The parameters characterizing plastic behavior of ballast have been defined based on a very conservative assumption taken from

$$F_l = 0.8F_p, \quad W_l = 0.02\ m. \tag{11}$$

Using the ballast parameters from *Table 2* maximum allowable temperatures of a freight train (axle load $F = 225\ kN$) have been determined for typical train velocities $V = 60, 90, 110, 130, 180\ km/h$ assuming a straight track with misalignments $\lambda = 5\ m$ and $\delta = 30\ mm$ (see *Figure 3*). They are plotted in *Figure 8*. From this figure it can be seen that the lateral resistance of ballast directly after renovation is poor so that if for instance actual equivalent temperature is

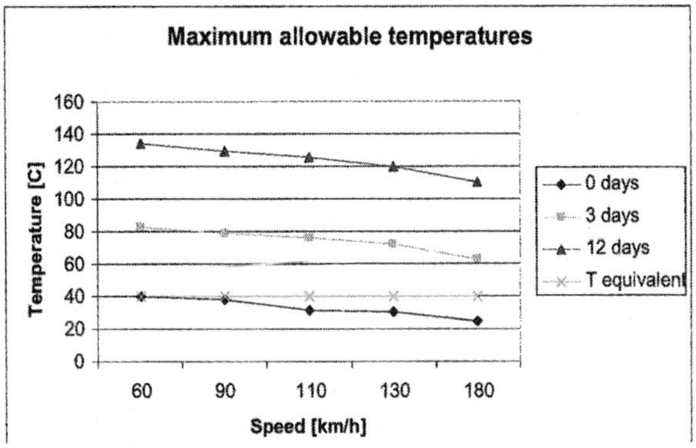

Figure 8. Maximum allowable temperature of rails for track after renovation

$T_{equiv} = 40°C$ (that corresponds to air temperature of $T_{air} = 21°C$ and no clouds) the maximum allowable speed of train is only $V^* = 60\ km/h$. Large improvement of ballast quality after 3 days can be explained by the use of ballast stabilization directly after the track maintenance.

Table 2. Ballast parameters after track renewing

Measurements			Ballast parameters			
	$\overline{F}_p^{ex}[N]$	$\overline{W}_p^{ex}[m]$	$F_p[N/m]$	$W_p[mm]$	$F_l[N/m]$	$W_l[mm]$
0 days	76500	0.0011	9993	0.17	7994	20
3 days	90260	0.0011	26584	0.35	21267	20
12 days	120030	0.0014	55643	0.69	44514	20

7. CONCLUSIONS

A procedure for identification of ballast parameters using measurement data of a real tack and numerical optimization technique has been presented. The procedure has been verified using measurement results with and without extra vertical load.

A technique for determination of speed limits based on UIC safety criteria has been described. The technique can provide a ground for assignment of temporary speed limits on tracks after renovation or after full maintenance.

References

1. ERRI-D202 (1996) Laying and Maintenance of CWR Track. International Union of Railways Leaflet 720 R, October 1996.
2. Hald, J, Madsen, K. (1985) Combined LP and Quasi-Newton Methods for Nonlinear L1 Optimization. SIAM J. Numer. Anal., Vol. 20, 68-80.
3. Markine, V.L., Esveld, C. (1999) Analysis of Lateral Behaviour of a Railway Track Structure Using an Optimization Technique. *Report 7-99-124-1, Laboratory for Road and Railway Engineering, Delft University of Technology*. ISSN 0169-9288
4. NS/CTO (1990) Zijdelingse Weerstandsmetingen van Drie Proefvakken een Referentievak van Elk 300 m op het traject Sittard-Roermond km 22.7-23.9 uitgevoerd in week 44 1989. *Rapport CTO/6/10.256/010*. (in Dutch)
5. NS/CTO (1991) Zijdelinkse weerstandmetingen op het traject Weert-Roermond km 74.430-78.250 uitgevoerd in week 35 van 1990, betreffende een nieuwe werkwijze van dwarsliggers verwisselen met behulp van de Mechanische Dwarsligger Verwisseling (MDV). *Rapport CTO/6/10.473/0005*. (in Dutch)
6. Samavedam, G. (1995) *Theory of CWR Track Stability*. Foster-Miller, Inc. Walthman.
7. Van, M.A. (1997) *Stability of Continuous Welded Rail Track*. Ph.D. Thesis, Delft University Press. ISBN 90-407-1485-1

ESTIMATION OF MATERIAL PARAMETERS AT ELEVATED TEMPERATURES BY INVERSE MODELLING OF A GLEEBLE EXPERIMENT

M. Eriksson, B. Wikman and G. Bergman
Department of Mechanical Engineering
Luleå University of Technology
SE-971 87 Luleå, Sweden

Keywords: Inverse modelling, thermomechanical testing, finite element analysis

Abstract A method for estimation of inelastic material parameters from Gleeble experiments is presented. Material parameters at elevated temperatures are estimated by inverse modelling of a Gleeble compression test during continuous cooling. The direct problem is solved using the implicit finite element program NIKE2D. The material behaviour is described by a rate independent thermo-elastic-plastic model with a nonlinear isotropic hardening law. The objective function is formed based on the discrepancy in force-displacement data between the numerical model prediction and the experiment. Minimisation of the objective function with respect to the material parameters is performed using an in-house optimisation software shell which is built on the subplex method. A Gleeble experiment with different deformation and temperature histories has been used for validation of the computed parameters. The validation experiment is simulated using both the material parameters achieved through inverse modelling and traditional estimation methods. The analysis showed better agreement when using the parameters estimated by the numerical method presented in this paper.

1. INTRODUCTION

Simultaneous forming and quenching is a manufacturing process for high-strength automotive steel components. The performance of these components is greatly influenced by the microstructure, residual stresses and geometrical defects developed during the process. The capacity of

P. Ståhle and K.G. Sundin (eds.), IUTAM Symposium on Field Analyses for Determination of Material Parameters - Experimental and Numerical Aspects, 151–165.

modern computers provide the means to simulate and analyse manufacturing processes by numerical methods such as the finite element (FE) method. The use of numerical simulations in the product development process have shown to reduce both lead time and cost [1]. However, FE simulations of coupled thermomechanical processes require not only efficient computational tools, but also accurate and appropriate material data. That is, the property data must be consistent with the conditions of the material in the combined forming and quenching process. Traditionally, the material is characterised by several tension or compression tests at elevated temperatures and different strain rates. This is often both time consuming and expensive. Moreover, compression tests at constant temperatures require a holding time which may have an influence on the microstructure of the steel and consequently on the stress-strain behaviour.

The present paper describes one way to reduce the number of experiments, and to avoid diffusional transformations, by using compression tests during continuous cooling. The material parameters are then estimated by means of inverse modelling [2].

2. EXPERIMENTAL PROCEDURES

In the manufacturing process of interest, the material is subjected to a temperature history of heating and subsequent cooling. Since the complete thermomechanical history of the material will influence its properties, relevant deformation and temperature histories must be used in the material testing. The material used in this work is a boron-bearing steel, BOLOC 02, manufactured and delivered by Swedish Steel AB. A Gleeble 1500 thermomechanical simulator with lengthwise strain control has been employed to create the required thermomechanical schedules. The experimental procedure followed is insertion of the cylindrical test specimen in a vacuum chamber, resistance heating to the austenite temperature and subsequent compression between tungsten carbide anvils during controlled cooling. Tantalum foils are used to prevent sticking between the anvils and the specimen, and graphite foils are used for lubrication. The temperature is measured using a Pt/PtRh-thermocouple welded on the surface of the specimen. The axial and radial displacements, and the axial compression force are also measured during the experiment.

Two different compression tests during continuous cooling have been performed. Experimental data from the first test is used as input to the inverse problem, see figures 1, 7 and 8. To obtain a cooling rate of 50°C/s, a small cylindrical specimen with diameter $\varnothing = 4.0$ mm and

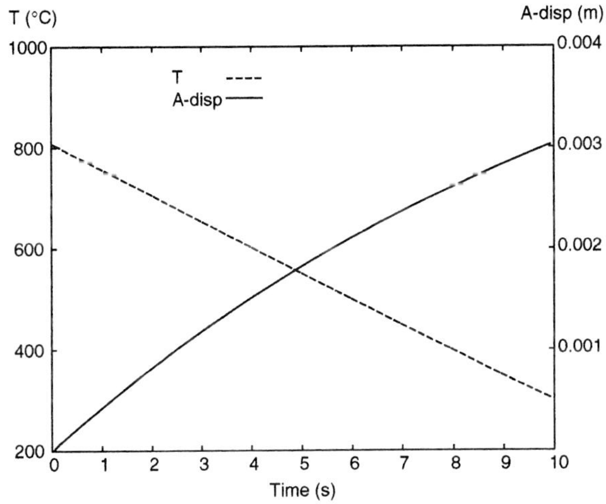

Figure 1 Temperature (T) and axial displacement (A-disp) (∅ = 4.0 mm, H = 5.5 mm).

height H = 5.5 mm is used. The compression starts at 800°C with a strain rate of 0.08 s^{-1} and ends at 300°C, resulting in an axial compression of 54%. For validation, the second test is performed with different strain and cooling rates, see figures 2, 9 and 10. This test specimen, ∅ = 5.0 mm and H = 7.0 mm, is compressed with a strain rate of 0.064 s^{-1} and a cooling rate of 40°C/s.

The elastic behaviour of the material is described by Young's modulus and Poisson's ratio. Poisson's ratio is considered to be constant, equal to 0.3. Young's modulus and thermal dilatation, obtained from [3], are shown in Figure 3. The thermal dilatation includes both thermal expansion and volume change due to phase transformation. The martensite start and end temperature are 420°C and 280°C, respectively.

For evaluation of the material parameters obtained from inverse modelling, stress-strain curves from isothermal compression tests assuming homogeneous strains have been used, see Figure 4. It should be noted that this data, which is presented in [3], include some effects from bainite formation caused by too slow cooling and an accelerated transformation due to deformation.

3. NUMERICAL PROCEDURES

The physical behaviour of the mechanical system considered is described by mathematical models. In addition, mathematical methods

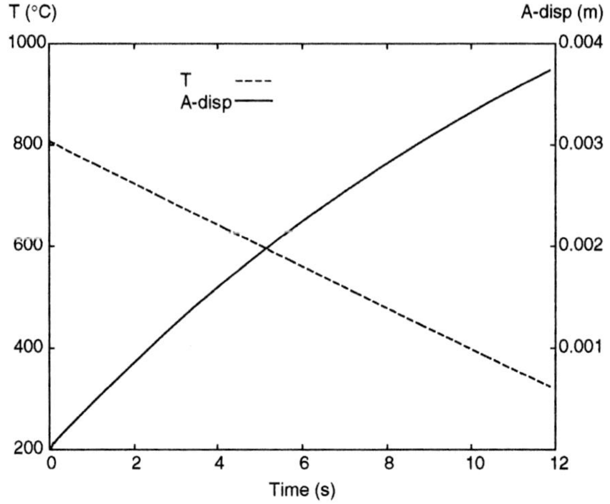

Figure 2 Temperature (T) and axial displacement (A-disp) (⌀ = 5.0 mm, H = 7.0 mm).

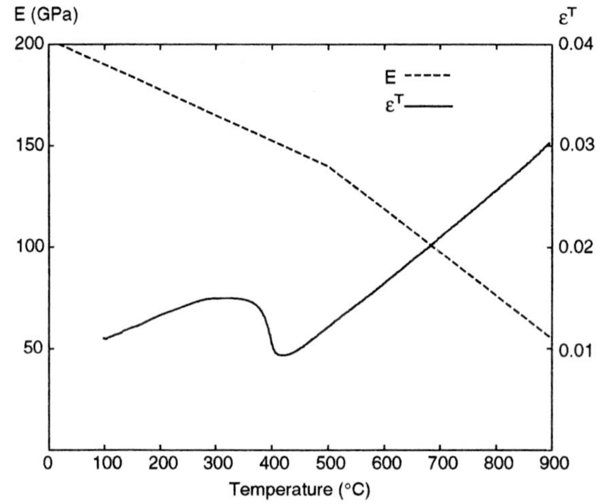

Figure 3 Young's modulus (E) and thermal dilatation (ϵ^T).

are used to study processes that are evolving in the material media and to estimate the necessary parameters. Here, the FE method is applied to solve the equations of motion of the direct problem and an optimisation algorithm is used to find the suitable material parameters. The approach is to simply repeat the computation with different material

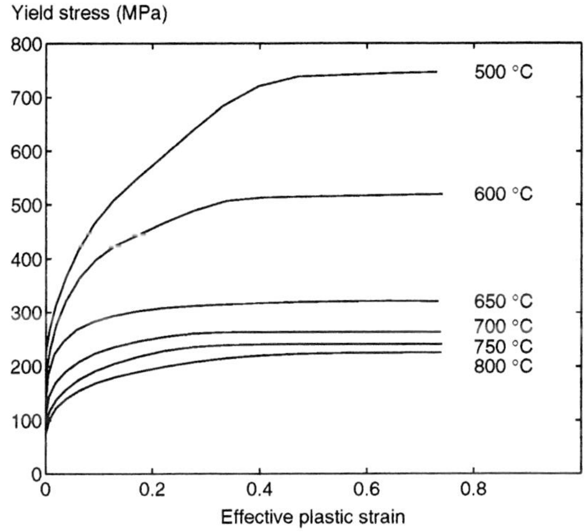

Figure 4 Traditionally estimated stress-strain curves.

characteristics, provided by the optimisation routine, until acceptable agreement with the experiment is obtained.

3.1. THE DIRECT PROBLEM

The direct problem consists of determining the response of the FE model which is subjected to variations of the material parameters. In the present work the implicit FE program NIKE2D [4] is used. The material model employed is a rate independent thermo-elastic-plastic model [5]. The plastic behaviour of the material is described by a von Mises isotropic yield condition, an associated flow rule and a nonlinear isotropic hardening law defined as tabulated values of yield stress versus effective plastic strain.

The axisymmetric FE mesh used in the optimisation problem of the Gleeble experiment is shown in Figure 5. The mesh, which is reduced to a quarter of the real geometry owing to symmetry conditions, consists of 59 four-node quadrilateral elements and 87 nodes. Displacement boundary conditions are applied to model the motion of the anvils, see Figure 1. The anvils are here considered to be rigid. Contact between the anvils and the test specimen is included in the model and the friction is assumed to follow Coulomb's friction law with a friction coefficient $\mu = 0.2$. The homogeneous temperature of the test specimen is prescribed according to the temperature history in Figure 1. The time step size used in all analyses is 0.005 s. A similar FE model is used to simulate the validation test.

The constitutive model utilises a set of yield stress versus effective plastic strain curves at different temperatures as input data. To reduce the number of unknown material parameters, we apply the following temperature (T) dependent relation between yield stress (σ_y) and effective plastic strain ($\bar{\epsilon}^p$)

$$\sigma_y = A(T)\left(B + C\bar{\epsilon}^{p^{1/n(T)}}\right) \tag{1}$$

in which

$$A(T) = a_0 + a_1T + a_2T^2 + a_3T^3 \tag{2}$$

and

$$n(T) = n_0 + n_1T \tag{3}$$

where a_0, a_1, a_2, a_3, B, C, n_0 and n_1 are the unknown parameters to be determined. It should be noted that in each time step, equations (1-3) are used to compute a stress-strain curve. The solution of the optimisation problem is more likely to converge to the global minimum if the initial parameters are chosen close to the optimum ones. The initial parameter values listed in Table 1 are determined from evaluations of Equation (1) with reference data from Figure 4. The elastic parameters, Young's modulus and Poisson's ratio, and the thermal dilatation are predetermined, see section 2.

3.2. THE INVERSE PROBLEM

The solution of the direct problem provides fundamental input information to the inverse problem. In the Gleeble application, the FE simulation renders input data in the form of axial reaction force and radial displacement histories as a function of the material model parameters,

$$D_i \Rightarrow [f_j, r_j]^{fem} \tag{4}$$

where D_i, $i = 1 \ldots M$, are the design variables which belongs to a set X of feasible material parameters. The solution of the direct problem, as illustrated in Equation (4), can be considered as the mapping of X into another set Y, which contains reaction forces f_j and radial displacements r_j at $j = 1 \ldots N$ discrete instants in time. A third set Z incorporates the reference data taken from the Gleeble experiment,

$$[f_j, r_j]^{exp} \in Z\ . \tag{5}$$

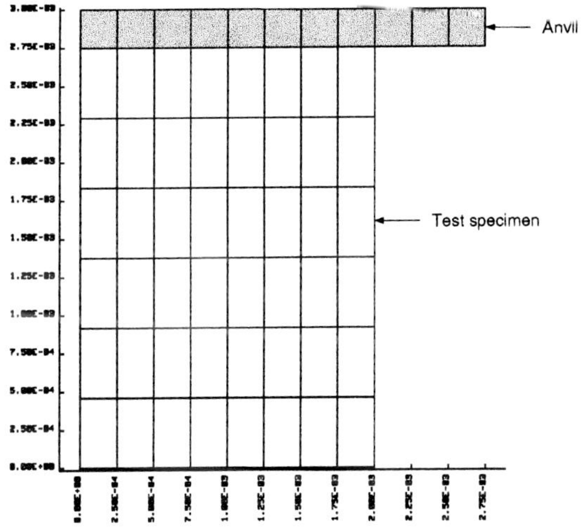

Figure 5 Axisymmetric FE mesh of the Gleeble experiment (Ø = 4.0 mm, H = 5.5 mm).

The objective of the inverse problem is to find the parameters D_i in X which give the best match between a solution in Y and the experimental reference data in Z. However, this procedure is not straightforward because in most cases Y and Z can not be matched exactly, i.e. $Y \neq Z$. Another obstacle is that even if the model would be able to exactly describe the real experimental behaviour, a reversal of Equation (4) is difficult. For this reason, an optimisation approach is utilised for estimating the material parameters which minimise the following objective function,

$$\varphi(D_i) = \frac{1}{2}\sum_{j=1}^{N}\left(\frac{f_j^{exp} - f_j^{fem}}{|f_{max}|}\right)^2 + \left(\frac{r_j^{exp} - r_j^{fem}}{|r_{max}|}\right)^2, \quad i = 1 \ldots M \qquad (6)$$

where $1/|f_{max}|$ and $1/|r_{max}|$ are weights that are applied to avoid adding terms of diverse orders of magnitude. The minimum of the objective function given in Equation (6) gives a closest match between a solution in Y and an experiment in Z, in a least-squares sense. The choice of criteria for the minimisation is an essential step in inverse problems. The least-squares form of parameter estimation can be considered as a type of maximum likelihood estimation, if the measurement errors are assumed to be independently random and normally distributed around the model prediction [6]. This approximation is often justified on the basis of the central limit theorem. That is, the sum of several different contributions will tend to be normally distributed, irrespective the probability distribution of the individual contribution, e.g. measurement errors.

In searching for the minimum value of the objective function φ, the subplex algorithm, a modified sequential simplex optimisation algorithm [7] has been used. This is a direct search method for unconstrained optimisation, i.e. the gradient of φ is not required. Usually, there are one or several constraints that must be satisfied in order to keep the solution in the feasible region of the parameter space. By adding penalty functions to the original objective function, the unconstrained optimisation method can be employed to solve the constrained problem. The direct search method used here allow for abrupt changes in the function value, such as step functions or similar. In the inverse modelling system used for the present estimation, the objective function value is doubled compared to the last successful iteration if one of the constraints is violated. Hence, the optimisation problem becomes,

$$min\ \varphi(D_i), \quad i = 1 \ldots M \tag{7}$$

subjected to

$$A(T)B \geq 1 \times 10^7 \tag{8}$$

$$A(T)C > 0 \tag{9}$$

$$0 < 1/n(T) \leq 1 \tag{10}$$

where the side constraints are given by equations (8-10). A basic flow chart of the inverse problem is depicted in Figure 6.

The inverse problem is solved by a tailor-made programming system [8]. It consists of a combination of analyser programs, interface processors and a connecting network. The analyser programs are in this case the FE, pre- and post-processing codes, and the optimisation program. Interface processors are short in-house programs for setting up the correct input format for each analyser program. The connecting network is the UNIX operating system which can be programmed through shell programs to control the data flow and to manage the necessary data communication. Errors which may occur in the FE analysis for some reason are taken care of by handling the error codes provided by the shell program. Erroneous FE runs are treated in the same way as the violation of a constraint.

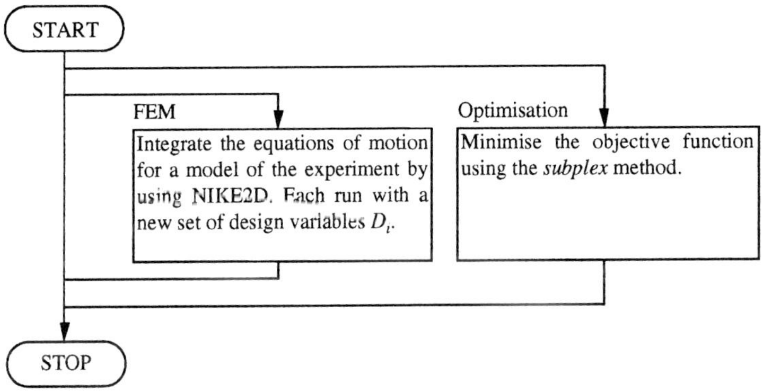

Figure 6 Basic flowchart of the inverse problem.

4. RESULTS

Results from the inverse analysis after 2000 iterations are presented in figures 7 and 8. Acceptable agreement between the computed and the experimental data is obtained. The corresponding material parameters are listed in Table 1. In the present estimation, the first 6 s of the experiment are considered. Consequently, the estimated material parameters are only valid in the temperature range 800°C to 500°C, see Figure 1. The reference values from the experiment are measured every 0.01 s, i.e. $N = 601$ in Equation (6). The complete inverse analysis took about 51 hours of CPU time on a Sun Ultra 10 440 MHz workstation.

Table 1 Initial and estimated material parameters.

Parameter	Initial value	Estimated value
a_0	-8.341×10^8	-7.09729×10^8
a_1	6.085×10^6	6.08653×10^6
a_2	-1.132×10^4	-1.13176×10^4
a_3	6.346	6.25813
B	1.259	1.11830
C	4.796	2.24423
n_0	-2.648	-2.67078
n_1	9.150×10^{-3}	1.19543×10^{-2}

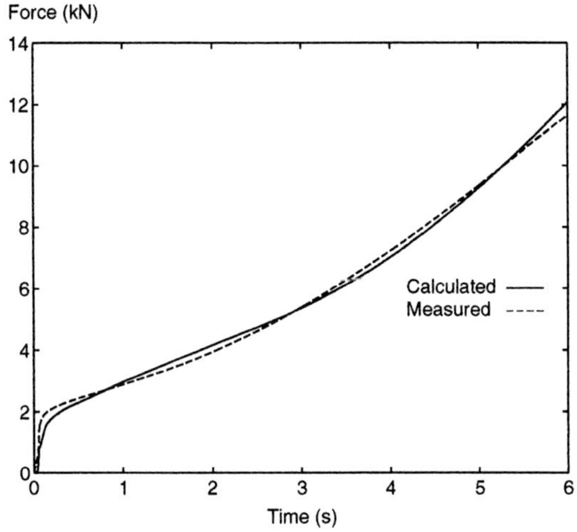

Figure 7 Measured and calculated axial force.

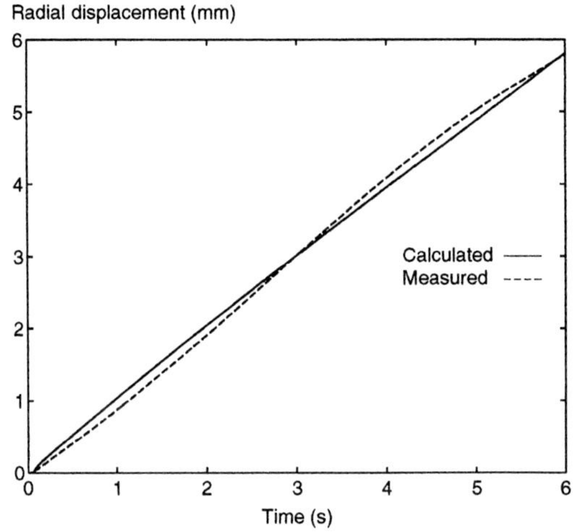

Figure 8 Measured and calculated radial displacement.

A friction coefficient $\mu = 0.2$ was used in the inverse analysis. To investigate the influence of friction on the estimated material parameters, four additional inverse analyses with different friction coefficients have been performed. Values of the objective function after 2000 iterations are listed in Table 2. It can be observed that the friction coefficient

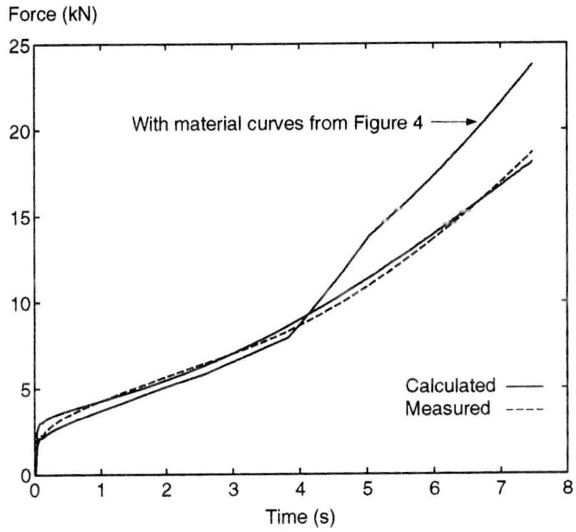

Figure 9 Measured and calculated axial force.

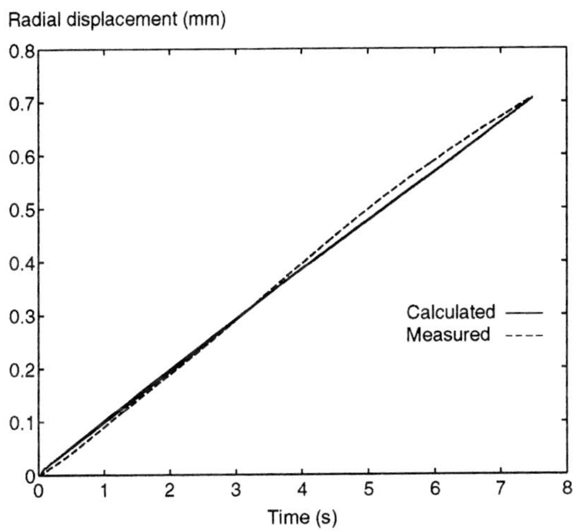

Figure 10 Measured and calculated radial displacement.

which corresponds to the best fit in axial force and radial displacement equals 0.2.

Results from the validation example are depicted in figures 9 and 10. The computed axial force is in reasonable agreement with the experimental result when using the material parameters achieved through inverse modelling. In the case of using the traditionally estimated curves, pre-

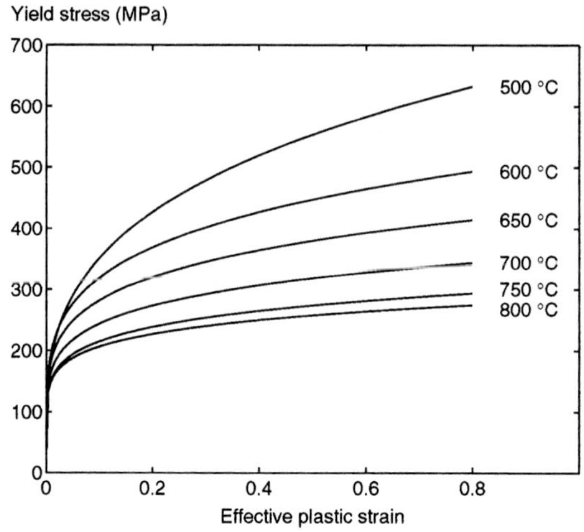

Figure 11 Stress-strain curves obtained by inverse modelling.

Table 2 Objective function values for different friction coefficients.

μ	$\varphi(D_i)$
0.10	0.61305
0.15	0.75764
0.20	0.38846
0.25	0.50753
0.30	0.41018

sented in Figure 4, the predicted force shows acceptable agreement with the measured force until t = 4 s, followed by an overestimation at sequential times. The radial displacements are almost identical in both simulations, and they are in reasonable agreement with the experiment.

For comparison with the material data in Figure 4, stress-strain curves based on equations (1-3) and the estimated parameters in Table 1 are shown in Figure 11. The curves in Figure 11 show slightly higher values for the yield stress in the temperature range 800°C to 650°C. Lower yield stress values are predicted at 600°C and 500°C.

5. DISCUSSION AND CONCLUSIONS

An inverse method for estimation of material parameters from a Gleeble compression test during continuous cooling is presented. Inelastic material parameters of a boron-bearing steel were determined in the temperature range 800°C to 500°C. The benefits of the method compared to traditional estimation methods are that the material can be characterised from a reduced set of experiments and that inhomogeneities in the stress and strain fields within the testpiece can be taken into account. The bulging of the test specimens does not impose any problem in the method presented here as long as the FE mesh can be adapted to the deformation. While the results of the present work are promising, there are a number of limitations and uncertainties which must be solved to improve the accuracy of the method. In particular, limitations and uncertainties related to the experimental data and the numerical methods used.

The incompleteness of the experimental data is believed to have a dominating effect on the accuracy of the estimated parameters. That is, the available data do not cover the whole range of deformation-temperature combinations needed for the model. To overcome this, additional experiments that provide combinations of low strains at low temperatures and high strains at high temperatures must be conducted. One optimisation cycle would then include the solution of several direct problems which all contribute to the objective function value. Furthermore, owing to the limitations of the current material model in describing diffusional transformations, only the first 6 s of the experiment has been used in the estimation. This deficiency can be avoided by using a more refined material model, which on the other hand most likely requires additional parameters and experiments. Another uncertainty is the friction coefficient between the anvil and the testpiece. Since the true value is unknown, the experimental data and the estimated parameters become functions of the friction coefficient. Consequently, the parameters reported here are those which corresponds to the best fit assuming that $\mu = 0.2$. To obtain a neutral set of parameters, the friction coefficient must be determined from a separate experiment.

Accuracy issues of the numerical methods together with systematic and random errors of the experimental data are always present. Numerical solutions of the thermomechanical problem and the optimisation problem are only approximate. In addition to the errors that might be introduced in the course of the development of the solution algorithm, or when setting up the boundary conditions, numerical solutions usually contain three kinds of systematic errors. Firstly, there are modelling errors, which can be defined as the difference between the true behaviour of the mechanical system and the exact solutions of the mathematical models. Secondly, the discretisation errors defined as the difference between the exact solution of the algebraic system of discretised equations and the originally continuous problem. Finally, the iteration errors which is the difference between the iterative and exact solution of the algebraic equation systems. The random component of the total measurement error can be estimated based on the repeatability error in the experimental data. The repeatability of the current experiment is qualitatively good but has not been quantified. A normal distribution of the readings will result even if some of the random error sources have distributions that are non-Gaussian, according to the central limit theorem [6]. The ultimate solution of the optimisation problem would be to reach an objective function value equal to zero. It would mean that an exact match is found, which usually is not possible in problems like this. Because of the nature of the problem, there can be more than one minimum of the objective function if different initial values of the design variables are used. In the current investigation, more than one minimum has been found. The solution presented here is the one which gave the lowest minimum among the solutions that have been computed.

Acknowledgments

The research funding by the European Community under the Brite-Euram programme, contract no. BRPR-CT97-0531, project no. BE97-4299, is gratefully acknowledged.

References

[1] Isaksson O., *Computational support in product development - Applications from high temperature design and development*, Doctoral thesis, 1998:46, Luleå University of Technology, Sweden, 1998.

[2] Wikman B., *Inverse modelling for determination of constitutive parameters - A literature review*, Technical report, 1999:21, Luleå University of Technology, Sweden, 1999.

[3] Eriksson M., *Modelling and simulation in the development of high strength steel components for vehicle structures*, Licentiate thesis, 2000:01, Luleå University of Technology, Sweden, 2000.

[4] Engelmann B. and Hallquist J. O., *NIKE2D - A nonlinear, implicit, two-dimensional finite element code for solid mechanics - User manual*, UCRL-MA-105413, Lawrence Livermore National Laboratory, USA, 1991.

[5] Bergman G. and Oldenburg M., *Verification of thermomechanical material models by thin-plate quenching simulations*, Journal of Thermal Stresses, vol. 20, 1997, pp. 679–695.

[6] Press W. H., Teukolsky S. A., Vetterling W. T. and B. P. Flanner, *Numerical recipes in FORTRAN*, Cambridge University Press, USA, 1992.

[7] Rowan T. H., *Functional stability analysis of numerical algorithms*, Doctoral thesis, The University of Texas at Austin, USA, 1990.

[8] Taylor D. L., *Computer aided design*, Addison-Wesley Publishing Company, USA, 1992.

HYDRIDE-INDUCED EMBRITTLEMENT IN METALS – STRESS AND TEMPERATURE EFFECTS

A.G. Varias
Materials Science, Technology and Society, Malmö University, SE-205 06 Malmö, Sweden

A.R. Massih
Quantum Technologies AB, Uppsala Science Park, SE-751 83, Uppsala, Sweden
Materials Science, Technology and Society, Malmö University, SE-205 06 Malmö, Sweden

Abstract A robust mathematical model for the hydrogen embrittlement of hydride forming metals has been developed. The model takes into account the coupling of the operating physical processes, namely: (i) hydrogen diffusion, (ii) hydride precipitation, (iii) non-mechanical energy flow, and (iv) hydride/solid-solution deformation. Crack growth is simulated by using a new version of de-cohesion model with time-dependent energy of de-cohesion due to the gradual process of hydride formation.

Zircaloy-2 hydrogen embrittlement and fracture initiation have been studied by using a finite element implementation of the model. Delayed hydride cracking has been considered in two configurations: (i) a semi-infinite crack, under mode-I K-field dominance and constant temperature, and (ii) a cracked plate, under tensile stress and temperature gradient. The initial and boundary conditions, in case (ii), are those encountered in the fuel cladding of boiling water reactors, during operation, and lead to loss of K-field dominance soon after the application of loading. The numerical simulation predicts hydride precipitation at some distance from the crack tip. The near-tip hydride platelets fracture, when the remote loading is sufficiently strong, and leave behind ligaments, which are stretched plastically, in agreement with experimental observations. The numerical results on hydride size, incubation period and crack growth velocity are compared with experimental data.

Further development of the model should be combined with accurate experimental determination of the mechanical and thermal properties of the hydrides.

1. INTRODUCTION

The stimulation of the present study is coming from nuclear industry, where zirconium alloys are used in several structural parts of a nuclear reactor core. Zirconium alloys combine good mechanical and neutron-absorption properties. However, hydrogen embrittlement occurs during service, mainly at extended fuel burnup (usage), and may lead to significant reduction in fracture toughness. Indeed, hydrogen is generated due to the oxidation of zirconium by the coolant water in the reactor. Subsequently hydrogen diffuses in the material and forms hydrides, when its terminal solid solubility is exceeded. The hydride is a brittle phase, which actually causes the embrittlement of the material. Then, delayed hydride cracking, a subcritical crack growth mechanism, may occur, e.g. [2]-[4], [7], [11].

P. Ståhle and K.G. Sundin (eds.), IUTAM Symposium on Field Analyses for Determination of Material Parameters - Experimental and Numerical Aspects, 167–177.

Hydrogen embrittlement, caused by the formation of hydrides at stress concentration locations, has been also observed in electron-microscope studies of vanadium, niobium and titanium [14], [1], [6], [12].

The hydride-induced embrittlement results from the simultaneous operation of several coupled processes, namely, (i) hydrogen diffusion, (ii) hydride precipitation, (iii) non-mechanical energy flow, and (iv) hydride/solid-solution deformation. A simulation of the embrittlement mechanism at constant temperature, i.e. by taking into account the coupling of processes (i), (ii) and (iv) has been presented recently by Lufrano, Sofronis and Birnbaum [8]-[9]. However, the heat transfer process within a nuclear reactor core leads to the development of temperature gradient in the fuel cladding and significantly affects material deterioration. A mathematical model, which takes into account all processes (i)-(iv) as well as hydride fracture for the simulation of crack growth, has been developed by Varias and Massih [15], [17]. The model is based on the thermodynamic theory of irreversible processes and takes into account hydrogen thermal transport. The time-dependent fracture toughness of the material ahead of the crack tip, due to the gradual hydride precipitation, is also taken into account in a new version of the de-cohesion model for crack growth. A brief discussion of the embrittlement/fracture model follows. Details of the model and relevant references are given in [15], [17].

2. HYDRIDE INDUCED EMBRITTLEMENT AND FRACTURE MODEL

The governing equations of all operating physical processes are presented. Hydrogen diffusion is simulated by enforcing the conservation of hydrogen mass:

$$\frac{dC^{HT}}{dt} = -\frac{\partial J_k^H}{\partial x_k}, \tag{1}$$

where C^{HT} is the total hydrogen concentration and J_i^H are the components of hydrogen flux. C^{HT} is related to the concentration of hydrogen in the solid solution, C^H, and the hydride, $C^{H,hr}$, as follows:

$$C^{HT} = fC^{H,hr} + (1-f)C^H, \tag{2}$$

where f is the hydride volume fraction. Note that C^H is equal to the hydrogen terminal solid solubility, C^{TS}, when $f \neq 0$, while $C^{H,hr}$ can be considered constant, independent of temperature. Also hydrogen diffusion in the hydride is significantly slower than in the metal and therefore it can be neglected. Then, the following relation provides the total hydrogen flux in a hydride/solid-solution composite:

$$J_k^H = -(1-f)\frac{D^H C^H}{R\mathrm{T}}\left(\frac{\partial \mu^H}{\partial x_k} + \frac{Q^H}{\mathrm{T}}\frac{\partial \mathrm{T}}{\partial x_k}\right). \tag{3}$$

Here R is the gas constant and T is the absolute temperature. Also D^H and Q^H are the diffusion coefficient and the heat of transport of hydrogen in the solid solution, respectively. Hydrogen thermal transport is taken into account by the temperature gradient term. The effects of stress and hydrogen concentration on diffusion are both included in the gradient of the chemical potential of hydrogen in the solid solution, μ^H:

$$\mu^H = \mu^{H,0} + \overline{V}^H\left(\frac{1}{2}M_{ijkl}\sigma_{ij}\sigma_{kl} - \frac{1}{3}\sigma_{mm}\right), \tag{4}$$

where $\mu^{H,0}$ is the stress free hydrogen chemical potential, which depends on hydrogen concentration. In the present study the law for ideal solutions is adopted. $\overline{V}^H$ is hydrogen molal volume. Also M_{ijkl} and σ_{ij} are metal's elastic compliance and stress tensor, respectively.

The precipitation of the hydride (MH_x) occurs under chemical equilibrium conditions, when the terminal solid solubility of hydrogen in the metal, C^{TS}, is reached:

$$C^{TS} = C^{TS,0}\exp\left(\frac{\overline{w}_{acc} + \overline{w}_{\mathrm{int}}}{\mathrm{x}RT}\right)\exp\left[\frac{\overline{V}^H}{RT}\left(\frac{\sigma_{mm}}{3} - \frac{1}{2}M_{ijkl}\sigma_{ij}\sigma_{kl}\right)\right], \tag{5}$$

where $C^{TS,0}$ is the terminal solid solubility of hydrogen, under stress-free conditions. Hydrides expand during precipitation and, consequently, the solubility of hydrogen is affected by the strain energy of hydride accommodation, $\overline{w}_{acc}$, as well as by the interaction energy of the applied stress field with the expanding hydride, $\overline{w}_{\mathrm{int}}$. Both $\overline{w}_{acc}$ and $\overline{w}_{\mathrm{int}}$ are defined per mole of hydride.

The non-mechanical energy flow is governed by the following differential equation, which is derived by enforcing conservation of energy:

$$\rho c_p \frac{d\mathrm{T}}{dt} + \frac{\Delta\overline{H}^{hr}}{\overline{V}^{hr}}\frac{df}{dt} = \frac{\partial}{\partial x_i}\left(k\frac{\partial \mathrm{T}}{\partial x_i}\right) - J_n^H\frac{\partial \mu^H}{\partial x_n}, \tag{6}$$

where ρ, c_p and k are the density, the specific heat at constant pressure and the thermal conductivity of the metal, respectively. Also $\Delta\overline{H}^{hr}$ and $\overline{V}^{hr}$ are the enthalpy associated with the formation of a mole of hydride and the hydride molal volume, respectively. Therefore the variation of the heat content in the metal–hydride composite depends on conducted heat, heat generated during hydrogen diffusion and heat released during hydride formation.

The deformation of the hydride/solid-solution composite is coupled with hydrogen diffusion and energy flow due to the strains, which are caused by hydrogen dissolution, hydride formation and thermal expansion:

$$\frac{d\sigma_{ij}}{dt} = M_{ijkl}^{-1}\left(\frac{d\varepsilon_{kl}}{dt} - \frac{d\varepsilon_{kl}^{H}}{dt} - \frac{d\varepsilon_{kl}^{E}}{dt}\right), \tag{7}$$

$$\frac{d\varepsilon_{ij}^{H}}{dt} = \frac{1}{3}\delta_{ij}\frac{d}{dt}\left[f\theta^{hr} + (1-f)C^{H}\overline{V}^{M}\theta^{H}\right], \tag{8}$$

$$\frac{d\varepsilon_{ij}^{E}}{dt} = \alpha\delta_{ij}\frac{d\Gamma}{dt}, \tag{9}$$

where θ^{hr} is the trace of hydride expansion strain tensor and $\theta^{H} = \overline{V}^{H}/\overline{V}^{M}$ is the expansion of the metal lattice due to hydrogen dissolution. $\overline{V}^{M}$ and α are the molal volume and the thermal expansion coefficient of the metal. It is assumed that the elastic and thermal properties of the hydride and the solid solution are identical.

Hydride fracture and therefore crack growth is simulated by considering cohesive tractions along the crack path. All information on material damage (i.e. void growth in the metal and hydride cleavage) is contained in the distribution of cohesive traction. At a material particle along the crack path, cohesive traction is related to the respective normal displacement through a simple trapezoid variation. The energy of de-cohesion, i.e. the energy required per unit crack advance, and the maximum cohesive traction are given by the following relations:

$$\phi_0 = f\phi_0^{hr} + (1-f)\phi_0^{M}, \tag{10}$$

$$\sigma_{\max} = \sqrt{f\sigma_{hr}^2 + (1-f)\sigma_M^2}\ . \tag{11}$$

Here ϕ_0^{M} and ϕ_0^{hr} are the de-cohesion energies, when there is no hydride and when there is only hydride, along the crack plane, respectively. ϕ_0^{hr} is derived from the threshold stress intensity factor of delayed hydride cracking, K_I^{hr}, [15]; $\phi_0^{hr} = (1-\nu^2)(K_I^{hr})^2/E$, where E, ν are Young's modulus and Poisson's ratio of the material. Similarly, ϕ_0^{M} is derived from the critical stress intensity factor of the metal, K_I^{M}. Also σ_{hr} is hydride fracture strength and σ_M is equal to three times the yield stress of the metal [15]. Note that the elastic-plastic behavior of the metal has been considered in the derivation of de-cohesion constitutive relations, in order to accurately predict the strong effect of hydrostatic stress on hydrogen diffusion.

In the following the hydrogen embrittlement and fracture model is applied to irradiated Zircaloy-2. The material properties are given in Table 1.

Table 1. Material properties used in the finite element calculations. The material properties correspond to irradiated Zircaloy-2 and δ-hydride ($ZrH_{1.66}$) [17]. C^{TS} is an experimentally derived terminal solid solubility under no applied stress.

E, ν	80.4 GPa, 0.369
σ_{hr}	580 MPa
σ_M	1740 MPa
K_I^M	30+0.045(T-300) MPa√m
K_I^{hr}	3.22+0.02205(T-300) MPa√m
D^H	$2.17\times10^{-7}\exp(-35087.06/RT)$ m²/s
Q^H	25122 J/mole
C^{TS}	$6.3741\times10^5\exp(-34542.75/RT)$ mole/m³
$C^{H,hr}$	1.02×10^5 mole/m³
$\overline{V}^H$	7×10^{-7} m³/mole
$\overline{V}^M$	14.0×10^{-6} m³/mole
$\overline{V}^{hr}$	16.3×10^{-6} m³/mole
θ^{hr}	0.1636
θ^H	0.05
x	1.66
ρ	6490 kg/m³
$\Delta\overline{H}^{hr}$	-63517.41 J/mole
k	9.37683+0.0118T W/mK
c_p	$226.69+0.206639T-6.4925\times10^{-5}T^2$ J/kgK
α	5.96×10^{-6} K⁻¹

3. HYDRIDE PRECIPITATION AND FRACTURE UNDER K-FIELD DOMINANCE

A semi-infinite crack is considered in a homogeneous material under mode-I loading and plane strain conditions. K-field traction is applied on a semi-circular boundary of radius equal to 0.1 m; only half of the space is analyzed, due to symmetry with respect to the crack line. The stress intensity factor initially increases at a rate equal to 0.2 MPa√m·s^{-1} for 100 s and subsequently remains constant at the maximum value of 20 MPa√m. The temperature of the material is equal to 300° C. Initial total hydrogen concentration is equal to 6438.5 mole/m^3 ($\approx$1000 ppm). The material is assumed to be insulated by a zirconium oxide layer and consequently zero hydrogen flux is prescribed along the crack face and the remote semi-circular boundary. Due to symmetry conditions, hydrogen flux is also taken equal to zero along the crack line. A Cartesian coordinate system is considered with origin at the crack tip and x_1-axis parallel to the crack plane and normal to the crack front.

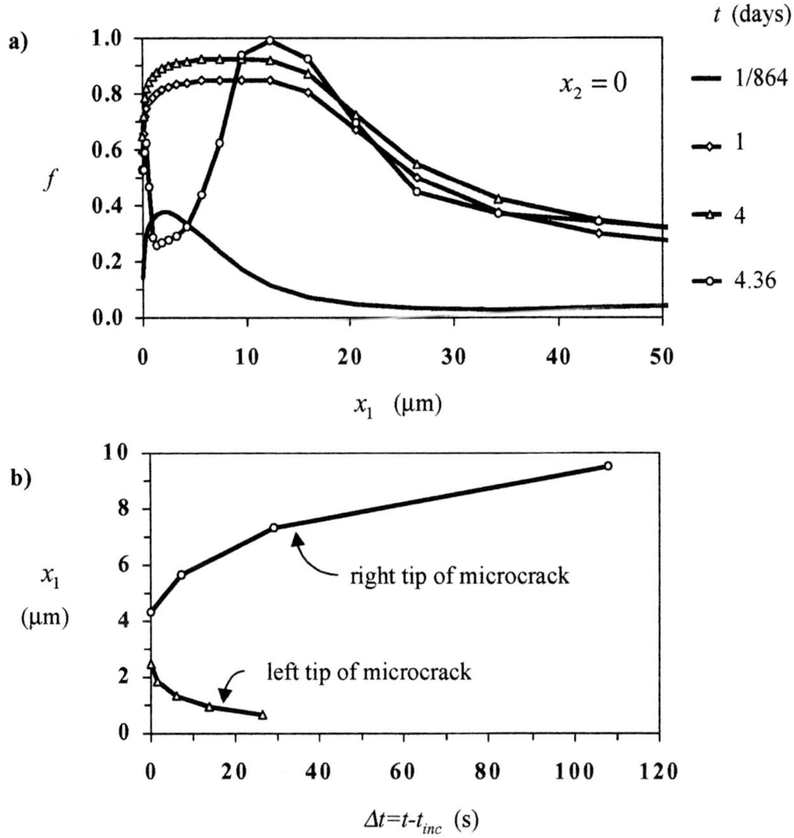

Fig. 1. **(a)** Hydride volume fraction distribution along the crack line and **(b)** microcrack propagation, under K-field dominance; t_{inc} =4.36 days.

The progress of hydride precipitation with time is presented in Fig. 1a. At t = 4 days, hydride volume fraction, f, is greater or equal to 0.90 within the range of 2.5 μm $\leq x_1 \leq$12.4 μm. The development of the area ahead of the crack tip, which is rich in hydride volume fraction, can be interpreted as the precipitation of near-tip hydrides of size about equal to 10 μm, in agreement with the interstriation spacing and the hydride platelet size, observed in delayed hydride cracking experiments [2], [7], [3]. At t = 4.36 days, the hydride volume fraction is relatively small ahead of the crack tip over a distance of the order of 10 μm. A new hydride volume fraction peak develops further away. Also there is a small region of about half a micron at the crack tip, where f is larger than 0.5. This redistribution of hydride volume fraction has been produced within about 100 s and it is attributed to the fracture of the near tip hydrides and the associated decrease of the near tip hydrostatic stress. Thus a micro-crack has been generated ahead of the main crack tip, leaving behind a thin ligament, which is expected to fail in a ductile mode, due to the relatively small hydride volume fraction

($0.5<f<0.65$). Therefore the present numerical simulation is in agreement with the development of striations during delayed hydride cracking (e.g. [2], [3]). According to Fig. 1b, the generated micro-crack propagates 8.88 μm, within the time period of 107.8 s, at an average speed of 8.24×10^{-8} m/s. Note that the micro-crack average speed does not correspond to the crack growth rate of stage II delayed hydride cracking, since we have simulated only the first cycle of the crack growth process. However, it is expected, that a variation of the average micro-crack speed, due to different material parameters, corresponds to a similar variation of the stage II crack growth rate.

The above calculations have been performed for K_I^{hr} equal to 9.24 MPa√m, which is an experimentally derived threshold stress intensity factor for delayed hydride cracking of irradiated Zircaloy-2. According to the experimental data of Huang and Mills [7], the threshold stress intensity factor for delayed hydride cracking decreases when the density of the hydride habit planes in the crack plane increases. Material texture affects also stage II crack growth rates. The crack growth rate increases with the density of the hydride habit planes in the crack plane. Therefore, by varying K_I^{hr} in the present model, texture effects could be implicitly taken into account. Indeed, crack growth calculations with $K_I^{hr}=1.63$ MPa√m show that the average micro-crack speed is doubled. The above threshold stress intensity factor is derived without the consideration of matrix plastic deformation, by using a cohesive zone model [16].

Efsing and Pettersson [3] measured, recently, a stage II crack growth rate equal to 9.5×10^{-7} m/s; the irradiated Zircaloy-2 specimens had initial hydrogen concentration between 560 and 1900 ppm and were tested at 300° C. Also, the incubation period varied between 16 and 24 hr. However, before testing, the specimens were subjected to a thermal cycle in order to promote crack growth; the specimens were kept at a higher temperature for a few hours, before the application of the loading. According to Simpson and Puls [13], when specimens are cooled to the test temperature the relatively small hydrides that form and grow during this excursion are more susceptible to dissolution, due to high elastic constraint, and this enhances the driving force for hydrogen diffusion into the crack-tip process zone. Note that, in the present simulation, the initial hydrogen concentration is assumed to be uniform, associated with an unconstrained material expansion, which does not lead to the development of residual stresses. The above mentioned temperature effect on crack growth could be introduced by considering a non-uniform initial hydrogen concentration which should be also associated with an initial non-zero and non-uniform stress distribution.

4. HYDROGEN EMBRITTLEMENT IN A CRACKED PLATE UNDER NO K-FIELD DOMINANCE

The effect of surface cracks on hydrogen embrittlement in the cladding of light water reactor fuel rods is investigated. The most critical case corresponds to a crack along the rod axis. Indeed, during reactor operation, the fuel pellet-cladding mechanical interaction is the major source of loading and may lead to the development of significant tensile hoop stresses [10]. The expansion of zirconium oxide on the waterside of the cladding may also produce tensile hoop stresses.

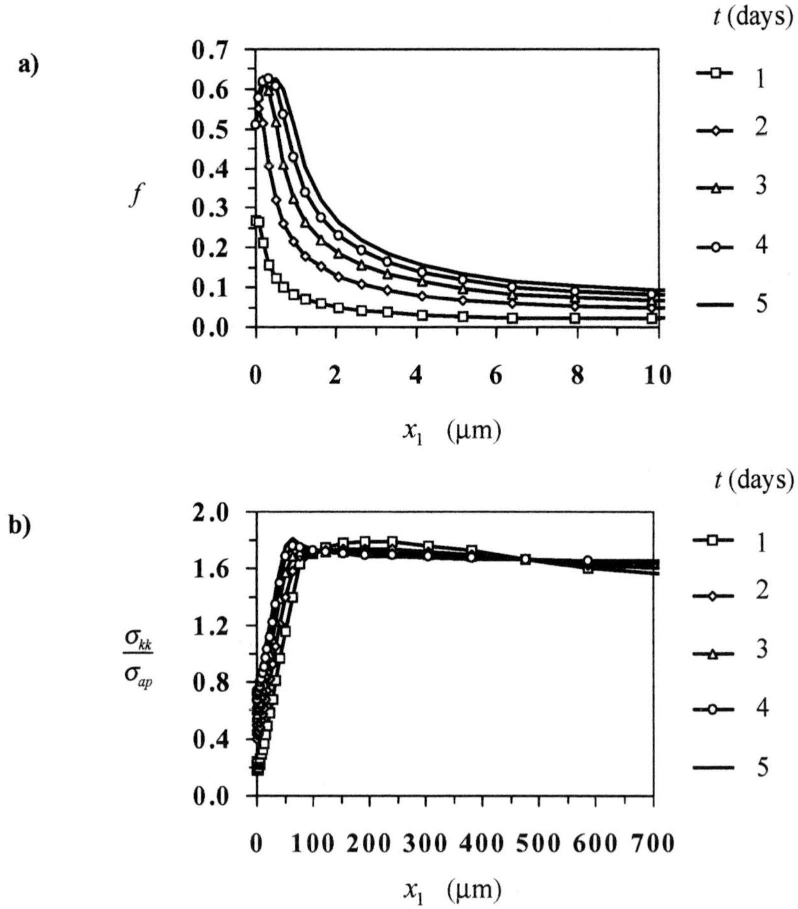

Fig. 2. **(a)** Hydride volume fraction and **(b)** stress trace distributions along the crack plane in a plate under tensile loading and temperature gradient.

A cracked plate is a good approximation of a cladding tube with an axial crack. The plate thickness, w, is taken to be equal to the actual cladding wall thickness (in this case w=0.8 mm). A long planar surface crack, of depth a (= 0.1w), has been considered on one side, which corresponds to the waterside of the cladding. A remote tensile stress, σ_{ap}, is applied normal to the crack faces and builds up gradually, at a rate of 10 MPa/s, up to the maximum value of 500 MPa. Zirconium oxidation is assumed to produce a constant inflow of hydrogen on the crack-side surface of the plate and the crack faces, which is equal to 0.122×10^{-7} mole/(m$^2\cdot$s) [5]. A zero hydrogen flux is considered on the other side of the plate, where stress and temperature gradients do not allow hydrogen outflow. The surface temperature is constant, being equal to 567 K, on the crack side, and 607 K, on the other side (i.e. the fuel side) [5]. The initial temperature distribution is assumed to be linear across the thickness of the plate. The initial hydrogen concentration is equal to 2500 mole/m^3 (≈388 ppm), which corresponds to 1900 days of reactor operation. Symmetry conditions require that the heat and hydrogen fluxes on the crack plane as well as on the remote plane, where σ_{ap} is applied, are zero. Plane strain conditions prevail. A Cartesian coordinate system is considered with origin at the crack tip and x_1-axis parallel to the crack plane.

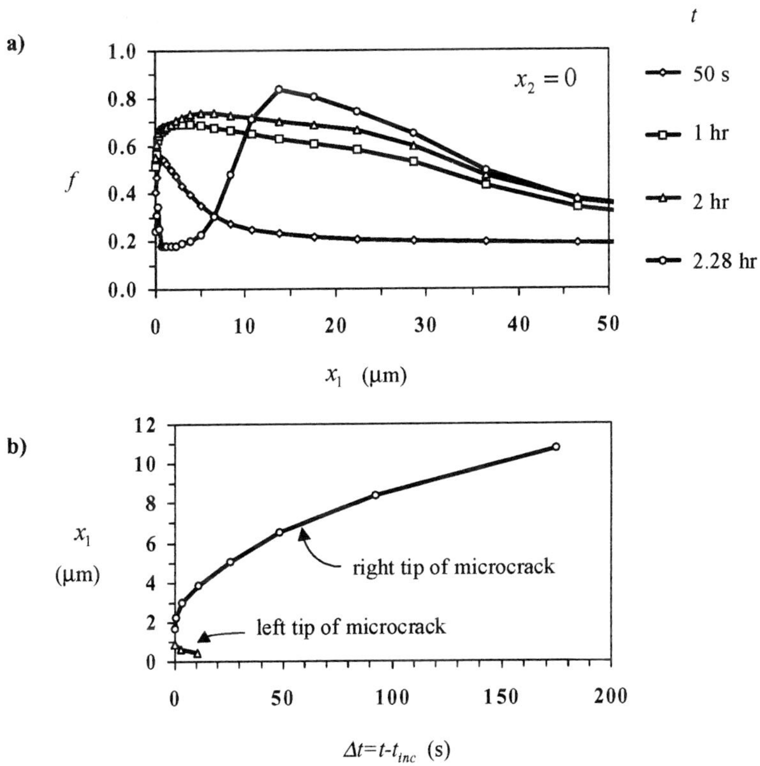

Fig. 3. **(a)** Hydride volume fraction distribution along the crack line and **(b)** microcrack propagation, in a cracked plate under tensile stress and temperature gradient; C^{HT} (t=0)= 20000 mole/m^3, t_{inc} =2.28 hr.

Figures 2a and 2b show the distribution of hydride volume fraction and normalized stress trace along the crack plane, respectively. Note that after 5 days of load application steady-state conditions are approached. According to Fig. 2b, there is no region of K-field dominance, due to hydrogen thermal transport to the waterside of the cladding. Therefore, linear elastic fracture mechanics cannot be used in integrity calculations. However, according to Fig. 2a, severe material embrittlement is limited only to a very small area near the crack tip and no crack growth is expected in the case under consideration. Indeed the finite element calculations show no unloading along the crack plane.

The numerical calculations were repeated for a crack length equal to a quarter of the cladding wall thickness (a=0.25w) as well as for a Zircaloy-2 toughness, which is by 25% smaller than the value given in Table 1; the selection of the new value of the fracture toughness is in agreement with the scatter in the experimental data. Four cases of initial hydrogen concentration were considered: 2500, 6438.5, 9657.7 and 20000 mole/m^3, which are about equal to 388, 1000, 1500 and 3200 ppm, respectively. Crack growth initiation was only predicted for the extreme case of 3200 ppm. Figure 3 shows the development of the near-tip hydride region and the respective microcrack propagation. The microcrack propagated 10.34 μm over a period of 174.80 s at an average speed of 5.91×10^{-8} m/s. The extreme hydrogen concentration was actually observed in places of the fuel cladding, where the oxide layer was partially lost due to spalling, during service. In this case, according to experiments, crack growth may occur.

Acknowledgments

A.G. Varias acknowledges the support of ABB Atom AB under the contracts 4500026874 and 4500028097. Financial support for the study on hydride induced embrittlement in metals is currently provided by the Foundation for Knowledge and Competence Development (Project Grant Hög 2000, KK-Stiftelsen, Sweden).

References

1. Birnbaum, H.K., Grossbeck, M.L. and Amano, M. Hydride precipitation in Nb and some properties of NbH, *Journal of the Less-Common Metals* **49** (1976) 357-370.

2. Dutton, R., Nuttall, K., Puls, M.P. and Simpson, L.A. Mechanisms of hydrogen induced delayed cracking in hydride forming materials, *Metallurgical Transactions A* **8** (1977) 1553-1562.

3. Efsing, P. and Pettersson, K. Delayed hydride cracking in irradiated Zircaloy cladding, in *Zirconium in the Nuclear Industry: Twelfth International Symposium, ASTM STP 1354*, edited by Sabol, G.P. and Moan, G.D., American Society for Testing and Materials, 2000, pp. 340-355.

4. Efsing, P. and Pettersson, K. The influence of temperature and yield strength on delayed hydride cracking in hydrided Zircaloy-2, in *Zirconium in the Nuclear Industry: Eleventh International Symposium, ASTM STP 1295*, edited by Bradley, E.R. and Sabol, G.P., American Society for Testing and Materials, 1996, pp. 394-404.

5. Forsberg, K. and Massih, A.R. Redistribution of hydrogen in Zircaloy, *Journal of Nuclear Materials* **172** (1990) 130-134.

6. Grossbeck, M.L. and Birnbaum, H.K. Low temperature hydrogen embrittlement of Niobium II – Microscopic observations, *Acta Metallurgica* **25** (1977) 135-147.

7. Huang, F.H. and Mills, W.J. Delayed hydride cracking behavior for Zircaloy-2 tubing, *Metallurgical Transactions A* **22** (1991) 2049-2060.

8. Lufrano, J., Sofronis, P. and Birnbaum, H.K. Elastoplastically accommodated hydride formation and embrittlement, *Journal of the Mechanics and Physics of Solids* **46** (1998) 1497-1520.

9. Lufrano, J., Sofronis, P. and Birnbaum, H.K. Modeling of hydrogen transport and elastically accommodated hydride formation near a crack tip, *Journal of the Mechanics and Physics of Solids* **44** (1996) 179-205.

10. Massih, A.R., Rajala, T. and Jernkvist, L.O. Analysis of pellet-cladding mechanical interaction behaviour of different ABB Atom fuel rod designs, *Nuclear Engineering and Design* **156** (1995) 383-391.

11. Sagat, S., Chow, C.K., Puls, M.P. and Coleman, C.E. Delayed hydride cracking in zirconium alloys in a temperature gradient, *Journal of Nuclear Materials* **279** (2000) 107-117.

12. Shih, D.S., Robertson, I.M. and Birnbaum, H.K. Hydrogen embrittlement of *a* titanium: in situ TEM studies, *Acta Metallurgica* **36** (1988) 111-124.

13. Simpson, L.A. and Puls, M.P. The effect of stress, temperature and hydrogen content on hydride-induced crack growth in Zr-2.5Pct Nb, *Metallurgical Transactions A* **10** (1979) 1093-1105.

14. Takano, S. and Suzuki, T. An electron-optical study of β-hydride and hydrogen embrittlement of vanadium, *Acta Metallurgica* **22** (1974) 265-274.

15. Varias, A.G. and Massih, A.R. Simulation of hydrogen embrittlement in zirconium alloys under stress and temperature gradients, *Journal of Nuclear Materials* **279** (2000) 273-285.

16. Varias, A.G. and Massih, A.R. Temperature and constraint effects on hydride fracture in zirconium alloys, *Engineering Fracture Mechanics* **65** (2000) 29-54.

17. Varias, A.G. and Massih, A.R. Hydride-induced embrittlement and fracture in metals – effect of stress and temperature distribution, *Journal of the Mechanics and Physics of Solids* (2002) in the press.

A NON-ISOTHERMAL MODEL FOR PHASE TRANSFORMATIONS AND THE MECHANICAL BEHAVIOUR OF A 12% CHROMIUM STEEL

Ralf Mohrmann
Fraunhofer-Institut für Werkstoffmechanik
D-79108 Freiburg, Wöhlerstrasse 11, Germany

Abstract A non-isothermal model was developed including viscoplastic deformation, phase transformations and their interactions. The model was applied to a 12% chromium steel for temperatures between RT and 1600K. Austenite transformation, martensite transformation and the constitution of carbides were identified to affect the mechanical behaviour of the martensitic steel investigated. A generalized Johnson-Mehl-Avrami equation and the Koistinen-Marburger equation were used to describe these phase transformations.
Viscoplastic deformations were modelled with an isothermal Chaboche-type model which was extended to non-isothermal conditions. Tensile, cyclic and relaxation test data were used to identify the model parameters.

1. INTRODUCTION

Finite element simulations of the weld fabrication process and of the post weld heat treatment require material models describing the mechanical behaviour and phase transformations. Those models should be applicable to multiaxial stress states, valid in the temperature domain considered, and should incorporate effects of phase transformations on the mechanical behaviour.

2. PHASE TRANSFORMATIONS IN STEEL X20CRMOV 12 1

2.1 Considered Phase Transformations

Phases and phase transformations of the 12% chromium steel X20CrMoV 12 1 during austenization and tempering were investigated experimentally by Petri et al. [1] and Straub [2]) and with thermodynamic calculations by Hald and Straub [3] and Strang [4]. In order to model the mechanical behaviour it is necessary to take at least those phases into account that affect the mechanical behaviour. The austenite phase, the martensite phase and a phase built by precipitated carbides were identified to affect the mechanical behaviour.

The relative mass proportions of the martensite phase z_M and of the austenite phase z_A (i.e. the mass proportion of metal atoms in a bcc lattice and in a fcc lattice,

P. Ståhle and K.G. Sundin (eds.), IUTAM Symposium on Field Analyses for Determination of Material Parameters - Experimental and Numerical Aspects, 179–187.

respectively) were defined. The unit cells of these phases carry different volumes. Therefore, a martensite or austenite transformation will cause a volume change.

The relative mass proportions of carbon dissolved in austenite c_A and in martensite c_M, or precipitated as carbides c_P were defined. For simplicity only one type of carbide was included. The volume change caused by the interchange of carbon was neglected.

2.2 Modelling Phase Transformations in Steel X20CrMoV 12 1

The phases to be modelled (z_A, z_M, c_A, c_M and c_P) obey obvious conserv ation laws

$$z_A + z_M = 1 \text{ and } c_A + c_M + c_P = 1 . \tag{1}$$

Basically, one has to consider the exchange rate of the austenite ↔ martensite tranformation $\dot{z}_A$ and the exchange rates of the carbon kinetics $\dot{c}_A$, $\dot{c}_M$ and $\dot{c}_p$, which are defined by (with $\dot{c}_A+\dot{c}_M+\dot{c}_p=0$)

$$\dot{c}_M = \dot{c}_{M\leftrightarrow P} - \dot{c}_{M\leftrightarrow A}, \quad \dot{c}_A = \dot{c}_{A\leftrightarrow P} + \dot{c}_{M\leftrightarrow A}, \quad \dot{c}_P = -(\dot{c}_{M\leftrightarrow P} + \dot{c}_{A\leftrightarrow P}) \tag{2}$$

$\dot{c}_{M\leftrightarrow P}$ and $\dot{c}_{A\leftrightarrow P}$ represent the rate of carbon that is interchanging between the precipitated and the dissolved state (in the martensite lattice or the austenite lattice). $\dot{c}_{M\leftrightarrow A}$ represents the amount of soluted carbon that is being e xchanged between martensite and austenite (in the domain where these phases coexist).

Austenization and the processes represented by $\dot{c}_{M\leftrightarrow P}$ and $\dot{c}_{A\leftrightarrow P}$ are governed by nucleation and growth. Concepts that treat this process have been developed by Johnson, Mehl and Avrami ([5], [6]). The generalized form based on the approach of Fortunier and Leblond [7] was used, here outlined for z_A:

$$\dot{z}_A = \begin{cases} m\, f(T)\cdot(z_{A,eq}(T) - z_A)\left(\ln\dfrac{z_{A,eq}(T)}{z_{A,eq}(T) - z_A}\right)^{\frac{m-1}{m}} & \text{if } z_{A,eq}(T) \geq z_A \\ f(T)\cdot(z_{A,eq}(T) - z_A) & \text{else} \end{cases} \tag{3}$$

where m is a temperature independent parameter. The kinetic parameter f and the equilibrium proportion of austenite $z_{A,eq}$ are appropriate functions of the temperature.

The martensite transformation is instantaneous. According to Koistinen and Marburger [8] the proportion of martensite at temperatures below the martensite starting temperature M_S is a function of temperature. The au stenite proportion z_A is defined

$$z_A(T) = \exp\left(-a(M_S - T)\right) \text{ with } a \approx 1.10\cdot 10^{-2} \text{ 1/K}. \tag{4}$$

For the carbon dissolved in the austenite or martensite phase it was assumed that this carbon keeps dissolved during austenite or martensite transformation, respectively

$$\dot{c}_{M\leftrightarrow A} = \begin{cases} \dfrac{\dot{z}_A}{1-z_A} c_M & \text{für } \dot{z}_A \geq 0 \\ \dfrac{\dot{z}_A}{z_A} c_A & \text{für } \dot{z}_A < 0 \end{cases} \tag{5}$$

Further information about the parameter identification, the experimental data base and the phase transformations is given in Gantert [9].

2.3 Simulation of an Industrial Heat Treatment

The present model was verified by the simulation (see figure 1) of a typical heat treatment defined in the industrial standard [10] normalizing at 1050°C for _ h, followed by air cooling and tempering at 760°C for 1 h, followed by air cooling. Figure 1 gives the temperature-time cycle of the simulated heat treatment.

Any initial material state should lead finally to a high proportion of martensitic lattice (z_A small) and almost all carbon precipitated as carbides (c_P almost 1). The evolution of phases is shown in figure 1. Heating and cooling rates are 10°C/s. The entire procedure can be described by the material behaviour within certain time intervals (compare Fig. 1):

Section I: The temperature has not reached austenite starting temperature A_{c1}. The solute carbon precipitates partly.

Section II: 1050°C is reached. The material is completely austenized and carbides start to dissolve.

Section III: During cooling Ms is reached. Austenite transforms quickly into martensite. Therefore, the proportion of carbon dissolved in austenite turns into a high proportion of carbon dissolved in martensite.

Section IV: The tempering temperature of 760°C is reached. Carbon content precipitates almost completely as carbides.

Section V: The heat treatment is finished. The material consists of 3% of austenite and 96% of the carbon is precipitated as carbide.

3. THE MECHANICAL MODEL

3.1 Model Definition

The mechanical model is based on the work of Chaboche ([11], [12] and [13] The definition of the one-dimensional form of the model is given by:

$$\dot{\varepsilon}_{vp} = p\,\text{sgn}(\sigma - a) \quad \text{where} \quad p = \left\langle \frac{|\sigma - a| - k}{Z} \right\rangle^n \tag{6}$$

$$a = a_1 + a_2 \tag{7}$$

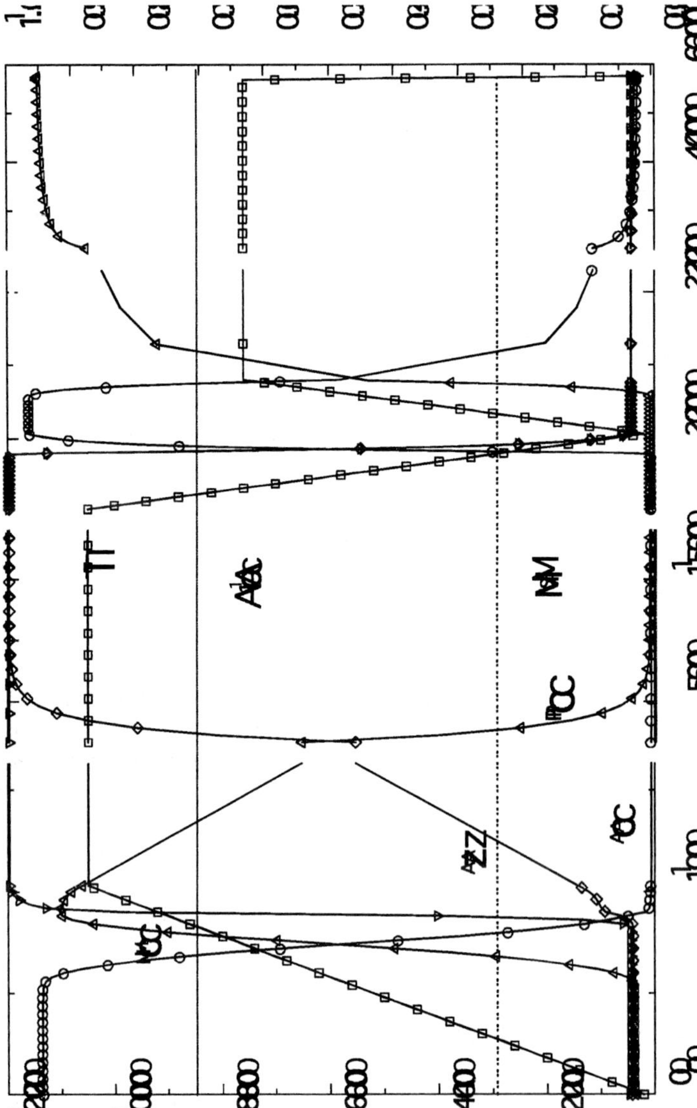

Fig. 1 Phase evolution during a complete heat treatment: temperature, and phase proportions are plotted versus time.

$$\dot{a}_i = H_i \dot{\varepsilon}_{vp} - D_i a_i \dot{p} \quad for \quad i = 1,2$$
$$a_i(t=0) = 0 \tag{8}$$

$$\dot{k} = 0, \qquad k(t=0) = k_0 \tag{9}$$

$\dot{\varepsilon}_{vp}$ is the viscoplastic strain rate. The kinematic hardening variables a_1 and a_2 and the isotropic hardening variable k are internal variables. The parameters for the yield stress, k_0, the stress exponent, n, the viscosity, Z, the cinematic hardening, H_1 and H_2, the dynamic recovery, D_1 and D_2, describe the behaviour for one temperature at a certain phase state. The mentioned parameter depend on temperature and phase variables (z_A, c_P, c_M and c_A). For every temperature and for every phase state mechanical data (tensile, tension/compression, relaxation tests) were needed to determine the parameters. It is obvious that this enormous experimental effort needs simplification. For the present work, mechanical data for three states:

tempered martensite ($z_A \approx 0$, $c_P \approx 1$, $c_A \approx c_M \approx 0$),

quenched martensite ($z_A \approx 0$, $c_M \approx 1$, $c_A \approx c_P \approx 0$) and

austenite ($z_A = 1$, $c_A = 1$, $c_P = c_M = 0$)

were available. The tempered martensite data were used to determine the parameters in the temperature range between room temperature and 800°C. Since there were few experiments performed in the state of high proportion of austenite only, following assumptions and simplifications were made arbitrarily, in order to reduce the flexability of the mechanical model:
Both, tempered martensite and austenite have individual yield stress parameter $k_{0,M}(T)$ and $k_{0,A}(T)$. It is assumed that they become identical at high temperatures.

A factor $b = \min\left(\dfrac{k_{0,A}(T)}{k_{0,M}(T)}, 1\right)$ is introduced. b is used to scale

$z_A(T) = b \cdot z_M(T)$ and $H_{i,A}(T) = b\ H_{i,M}(T), i = 1,2$.

All other austenite parameters are identical to the tempered martensite ones:

$n(T) = n_A(T) = n_M(T)$ and $D_i(T) = D_{i.A}(T) = D_{i,M}(T)$.

No plasticity can be detected from the quenched martensite data, because the material is brittle and fractures before yielding. We assume a quasi-elastic model by adding the term $c_M\ \Delta k_M$ (Δk_M = 1000 MPa) to the tempered martensite yield stress parameter $k_{0,M}(T)$.

The parameter values depending on the phase proportion are linearly interpolated with respect to the relative proportion of austenite z_A between the states of tempered martensite and austenite. The dependence of the isotropic hardening rate $\dot{k}$ and the initial value of k on the phase proportions are

$$\dot{k} = \left((1 - z_A)\frac{\partial k_{0,M}}{\partial T} + z_A \frac{\partial k_{0,A}}{\partial T} \right) \dot{T} + \left(k_{0,A}(T) - k_{0,M}(T)\right) \dot{z}_A + \Delta k_M \, \dot{c}_M \qquad (10)$$

$$k(t = 0) = (1 - z_A) k_{0,M}(T) + z_A \, k_{0,A}(T) + \Delta k_M \, c_M$$

The interpolations of the viscosity parameter Z and the hardening moduli H_i are given in the equations 11 and 12, respectively:

$$Z(T) = (1 - z_A) Z_M(T) + z_A Z_A(T) \qquad (11)$$

$$H_i(T) = (1 - z_A) H_{i,M}(T) + z_A H_{i,A}(T) \quad for \quad i = 1,2 \qquad (12)$$

Since the von-Mises criteria has been used, the model can easily be generalized to a multiaxial form and applied for future implementations in finite element programs.

3.2 Temperature Dependence of Model Parameters

The mechanical model was extended to non-isothermal conditions: First, the parameters for tempered martensite were determined for each temperature separately. Then, appropriate functions to approximate these parameters have to be identified. These functions depend on parameters themselves. Finally, these parameters are fitted to the whole set of given experiments.

Figure 2 illustrates this process for the parameters k, Z, and n. The symbols are the parameter values at RT, 200°C, 400°C, 600°C, 700°C, 800°C, 850°C, and 900°C. The lines are analytic functions fitted to these values. The dashed lines show the parameter values after fitting the parameter functions to all given experiments. These parameter values (dashed lines in Fig. 2) are used in the next section to discuss experimental and modelling results.

4. RESULTS

In figure 3 and 4 part of the experimental and modelling results are shown. Further results can be found in Gantert [9]. In figure 3, tensile tests with similar loading rates and different temperatures are depicted. The material shows the expected decreasing of yield stress for increasing temperature. The yield stress changes most at about 600°C which is in accordance with the parameter change in this temperature domain (see figure 2).

Furthermore, the tensile tests at 800°C, 850° and 900°C show virtually no difference, although one would expect, that a temperature difference of 100K and a change of the equilibrium proportion of austenite between 0 and 81% austenite should have an effect. The microstructures were equilibrated at the test temperature before the tensile tests were run. Figure 4 gives results for tension/compression tests. The model describes the experimental data reasonable well.

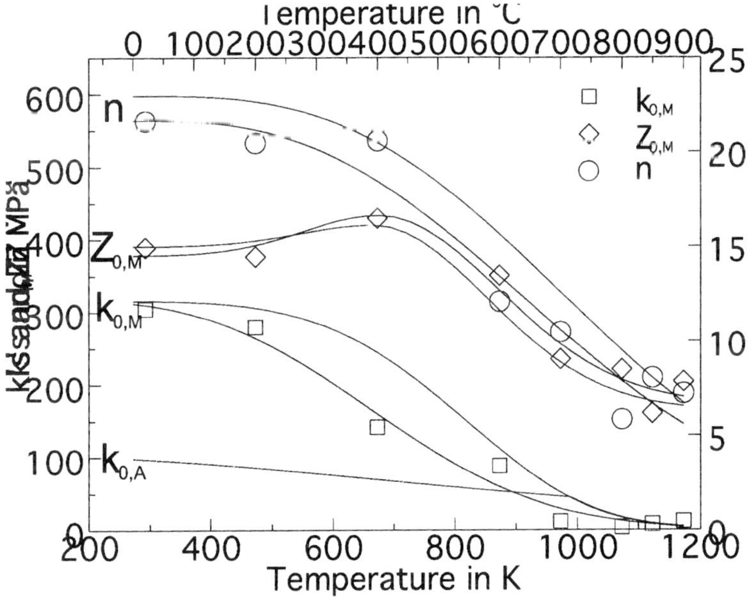

Fig. 2 Temperature dependence of model parameters: The symbols represent isothermal fits, the lines are analytic functions fitted to these parameters and the dashed lines are the parameter functions fitted to the whole set of experiments.

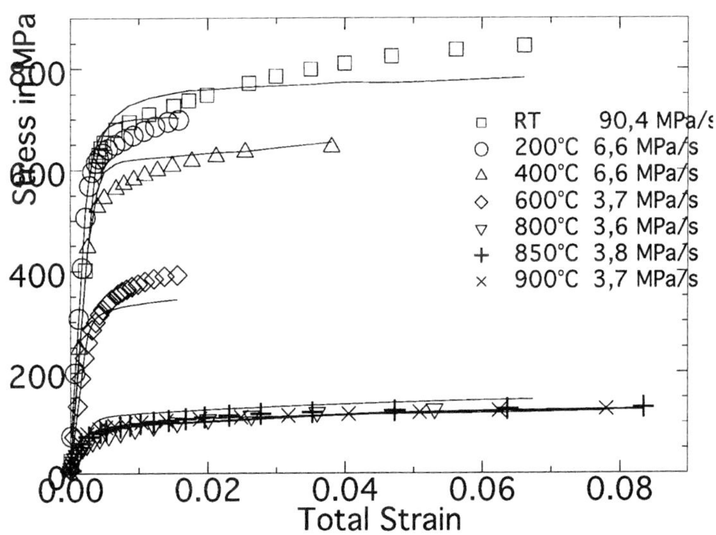

Fig. 3 Tensile test and modelling results for X20CrMoV 12 1: Stress is plotted versus total strain. Symbols represent experimental data, lines represent the corresponding model result.

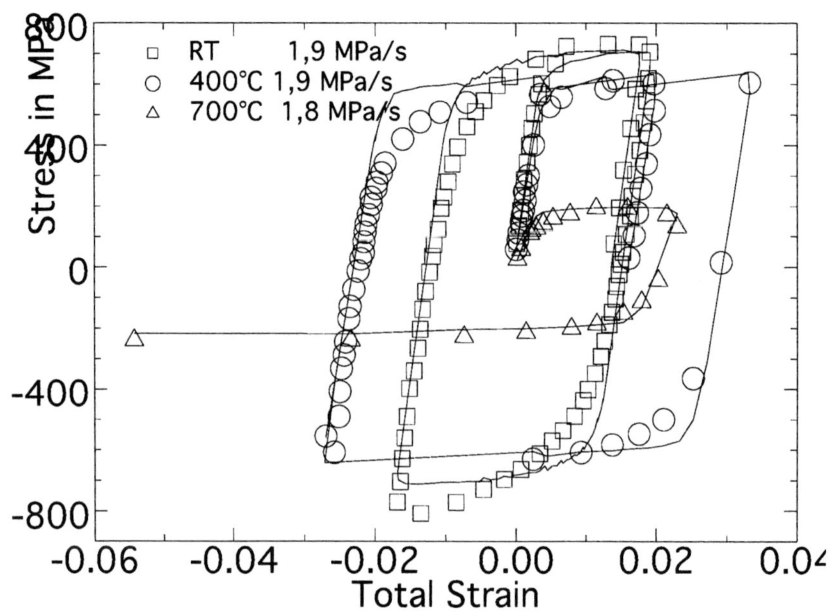

Fig. 4 Tension/compression tests and modelling results for X20CrMoV 12 1 at room temperature, 400°C and 700C°: All experiments start at zero stress and strain. Stress is plotted versus total strain. Symbols represent experimental data, lines represent the corresponding model result.

5. Conclusion

The presented material model is based on a Chaboche-type viscoplastic deformation model, a generalized Johnson-Mehl-Avrami equation describing austenite transformation and carbon kinetics and the Koistinen-Marburger equation describing martensite transformation.

The transformations between the tempered, quenched and austenized states of steel X20CrMoV 12 1 were modelled well. These transformations are relevant for the mechanical behaviour of X20CrMoV 12 1. The martensitic and the austenitic proportions of the material and their carbon content were modelled well. This was verified by a simulation of an industrial heat treatment.

Based on the transformation behaviour of steel X20CrMoV 12 1 the Chaboche model was fitted to mechanical data of austenite and martensite phases, respectively. In order to describe temperature dependent parameter functions were determined. For the martensitic and the austenitic state the model reproduces the experiments reasonably well.

Acknowledgement

We gratefully acknowledge the support of Dr. A. Rosselet, SULZER Innotec, Switzerland for supplying the huge data base and fruitful discussions.

References

[1] Petri, R., Schnabel, E., Schwaab, P., "Zum Legierungseinfluß auf die Umwandlungs- und Ausscheidungsvorgänge bei der Abkühlung warmfester Röhrenstähle nach dem Austenitisieren, II. 12%ige Chromstähle", Archiv des Eisenhüttenwesens 52 (1981) 27

[2] Straub, S., "Verformungsverhalten und Mikrostruktur warmfester martensitischer 12%-Chromstähle", Fortschrittsberichte VDI, Reihe 5: Grund- und Werkstoffe, Nr. 405, VDI Verlag, 1995

[3] Hald, J., Straub, S., "Microstructural Stability of 9%-12%CrMo(W)VNbN-Steels", preprint

[4] Strang, A.. (Hrsg.), "Microstructural Development and Stability in High Chromium Ferritic Steels", Article in Book "Microstructural development and stability in high chromium ferritic power plant steels", Cambridge University Press, 1997

[5] Johnson, W., A., Mehl, R., F., "Reaction Kinetics in Processes of Nucleation and Growth", Transactions of the American Institute of Mining, Metallurgical, and Petroleum Engineers 135 (1939) 416

[6] Avrami, M., Kinetics of Phase Change I, Journal of Chemical Physics 7 (1939) 1103

[7] Fortunier, R., Leblond, J.-B., "Recent advances in the numerical simulation of simultaneous diffusion and precipitation of chemical elements in steels", Framasoft+CSI, 10 rue Juliette Récamier, 69456 Lyon Cedex 06, France

[8] Koistinen, D., P., Marburger, R., E., Acta Metallurgica 7 (1959) 59

[9] Gantert, C., "Modelle für das temperaturabhängige Verformungsverhalten von Hochtemperaturwerkstoffen und Anwendung auf Kraftwerksstähle", 1999, Technical Report, Fraunhofer IWM, Freiburg, Germany

[10] DIN 17 175, Nahtlose Rohre aus warmfesten Stählen, 5/1979

[11] Chaboche, J.-L., Rousselier, G., "On the Plastic and Viscoplastic Constitutive Equations Part I: Rules Developed with Internal Variable Concept", Journal of Pressure Vessel Technology 105 (1983) 153

[12] Chaboche, J.-L., "Time-Independent Constitutive Theories for Cyclic Plasticity", International Journal of Plasticity 2 (1986) 149

[13] Chaboche, J.-L., Lemaitre, J., "Mechanics of Solid Materials", Cambridge Univ. Press, 1990

SIZE EFFECT ON STRESS-INTENSITY FACTORS OF CNS SPECIMEN WITH AN INTERFACE CRACK SUBJECTED TO MIXED-MODE LOADING

Kenji Machida
Department of Mechanical Engineering, Science, University of Tokyo
2641, Yamazaki, Noda-shi, Chiba, 2788510, JAPAN

Abstract. A three-dimensional finite element analysis is carried out by using a compact normal and shear (CNS) specimen with various kinds of specimen size. The complex stress intensity factor K associated with an elastic interface crack is evaluated by the virtual crack extension method. The effect of Young's modulus and Poisson's ratio on stress intensity factors is discussed under various kinds of mixed-mode loading. It is predicted that the stress intensity factors would vary with the specimen size. Then, the size effect on stress-intensity factors is discussed. A simple evaluation method for stress intensity factors is needed in the fracture toughness test. Therefore, a polynomial fit is proposed to evaluate the stress intensity factors at the midsection of a CNS specimen with an interface crack subjected to mixed mode loading. It is possible to evaluate the stress intensity factors of a CNS specimen with high accuracy by the present polynomial evaluation.

1. INTRODUCTION

In recent years, structures using different-species material, such as an IC packages, adhesive joints, electron devices, and composites, are being increasingly used. The strength evaluation of an interface has made mainly analytically and theoretically[(1)-(31)]. However, methods for testing have not yet been established for the quantitative strength-evaluation of the interface crack. The compact normal and shear (CNS) specimen proposed by Richard and Benitz [(32)] enables us to carry out the experiments under various kinds of mixed-mode loading. The evaluation of stress-intensity factors of the interface crack has been carried out by the conservation M-integral[(7)], a new method [(14)] combining the idea of the M-integral with the virtual crack extension method (VCEM)[(33)], the modified crack-closure integral[(7),(34),(35)], and the stress extrapolation[(29)(30)]. The high accuracy of the virtual crack extension method has been reported in previous papers [(29),(30)].

In this study, three-dimensional finite element analysis was carried out using the CNS specimen with various kinds of Young's modulus, Poisson's ratio and a load application angle. The stress-intensity factors were evaluated by the virtual crack extension method. The stress-intensity factors at the central part of the specimen are nearly uniform. The energy release rate of the interface crack varies with on the mixed-mode ratio. Therefore, the fracture toughness of the interface crack should be estimated by the components of the stress intensity factor ($K_{\rm I}$ and $K_{\rm II}$). It is predicted

P. Ståhle and K.G. Sundin (eds.), IUTAM Symposium on Field Analyses for Determination of Material Parameters - Experimental and Numerical Aspects, 189–202.

that the stress-intensity factors would vary with the specimen size. The size effect on stress-intensity factors is discussed. A simple evaluation method for stress-intensity factors is needed in the fracture toughness test. Therefore, the polynomial fit of stress-intensity factors was carried out on CNS specimen with various of Young's moduli, Poisson's ratio, load application angles and specimen sizes. The stress-intensity factors of CNS specimen can be simply evaluated by the present polynomial approximation.

2. STRESS-INTENSITY FACTORS OF INTERFACE CRACK

2.1 Complex stress-intensity factor

Erdogan[2] introduced the complex stress-intensity factor which characterizes the singular field as:

$$(\sigma_{yy} + i\sigma_{xy})_{\theta=0} = \frac{K(r/l_k)^{i\varepsilon}}{\sqrt{2\pi r}} \tag{1}$$

$$\varepsilon = \frac{1}{2\pi}\ln\left[\frac{\mu_1 + \mu_2\kappa_1}{\mu_1\kappa_2 + \mu_2}\right] \tag{2}$$

$$\kappa_j = \begin{cases} 3-4\nu_j & \text{plane stress} \\ (3-\nu_j)/(1+\nu_j) & \text{plane strain} \end{cases} \tag{3}$$

where $(\sigma_{yy} + i\ \sigma_{xy})$ is the complex stress, K is a complex stress-intensity factor, l_k is the typical length like a crack length used to set the dimension of the stress-intensity factor of the interface crack in correspondence to that of the homogeneous material, E_1 and E_2 are shear moduli of materials 1 and 2, r is the distance from the crack tip, ε is a bi-material constant, and ν is Poisson's ratio.

The absolute value and argument are expressed as follows:

$$\left|\sigma_{yy} + i\sigma_{xy}\right| = \frac{|K_{\mathrm{I}} + iK_{\mathrm{II}}|}{\sqrt{2\pi r}} = \frac{\sqrt{K_{\mathrm{I}}^2 + K_{\mathrm{II}}^2}}{\sqrt{2\pi r}} \tag{4}$$

$$\arg\left(\sigma_{yy} + i\sigma_{xy}\right) = \tan^{-1}\left(K_{\mathrm{II}}/K_{\mathrm{I}}\right) + \varepsilon\ln\left(r/l_k\right) \tag{5}$$

When l_k changes to l', K_{I}' and K_{II}' are simply evaluated by the following equation,

$$\begin{Bmatrix} K_{\mathrm{I}}' \\ K_{\mathrm{II}}' \end{Bmatrix} = \begin{bmatrix} \cos\theta & -\sin\theta \\ \sin\theta & \cos\theta \end{bmatrix} \begin{Bmatrix} K_{\mathrm{I}} \\ K_{\mathrm{II}} \end{Bmatrix}, \qquad \theta = \varepsilon\ln\left(\frac{l'}{l_k}\right) \tag{6}$$

2.2 Virtual crack extension method

The strain energy release rate in the virtual crack extension method proposed by Parks[33] is expressed as follows:

$$G=-\left(\frac{\partial U}{\partial A}\right)=-\frac{1}{2}\sum_{i=1}^{N_F}u_i^T\frac{\Delta k_i}{\Delta A}u_i \tag{7}$$

where U is the potential energy, ΔA is the increment of crack area, Δk_i is the variation in the stiffness matrix of elements (1-N_F) within an interior contour which surrounds the crack tip with ΔA, and u_i is the vector of nodal point displacement.

According to Yau-Wang-Matos[(6),(14)], the stress-intensity factor solution for the mixed-mode fracture problem can be obtained as follows. Consider two independent equilibrium states with field variables denoted by superscripts 1 and 2 within an interior contour Ω, which surrounds the crack tip. The stress-intensity factors of equilibrium state 2 are known. Denoting the superposed equilibrium state by a superscript 1+2, the principle of superposition for the mixed-mode stress-intensity factors holds as

$$\begin{aligned} K_{\mathrm{I}}^{(1+2)} &= K_{\mathrm{I}}^{(1)}+K_{\mathrm{I}}^{(2)} \quad K_{I}^{(1+2)} = K_{I}^{(1)}+K_{I}^{(2)} \\ K_{\mathrm{II}}^{(1+2)} &= K_{\mathrm{II}}^{(1)}+K_{\mathrm{II}}^{(2)} \quad K_{II}^{(1+2)} = K_{II}^{(1)}+K_{II}^{(2)} \end{aligned} \tag{8}$$

The relation between G and K_{I}, K_{II} is expressed as follows:

$$G=\beta\left(K_{\mathrm{I}}^2+K_{\mathrm{II}}^2\right) \tag{9}$$

$$\beta=\left(c_1+c_2\right)/16\cosh^2(\pi\varepsilon) \tag{10}$$

$$c_1=\left(\kappa_1+1\right)/\mu_1,\quad c_2=\left(\kappa_2+1\right)/\mu_2 \tag{11}$$

From eqs. (8), (9) and (10), the following equation can be obtained.

$$2\beta\left[K_{\mathrm{I}}^{(1)}K_{\mathrm{I}}^{(2)}+K_{\mathrm{II}}^{(1)}K_{\mathrm{II}}^{(2)}\right]=G^{(1+2)}-G^{(1)}-G^{(2)} \tag{12}$$

$G^{(2)}$ is obtained by substituting β, $K_{\mathrm{I}}^{(2)}$ and $K_{\mathrm{II}}^{(2)}$ into eq. (12). $G^{(1)}$ and $G^{(1+2)}$ can be obtained by substituting $u^{(1)}$ and $u^{(1+2)}$ into eq. (7). Therefore, $K_{\mathrm{I}}^{(1)}$ and $K_{\mathrm{II}}^{(1)}$ can be obtained from eq. (13) by superposing two known solutions on the analytical model.

Let us consider the superposition of the asymptotic solution of plane strain derived by Sun and Jih[(7)]. Superposing the plane strain solutions of $K_{\mathrm{I}}^{(2)}=1$, $K_{\mathrm{II}}^{(2)}=0$, and $K_{\mathrm{I}}^{(2)}=0$, $K_{\mathrm{II}}(^{2)}=1$, respectively, and substituting the superposed results into eq. (12), we can obtain $K_{\mathrm{I}}^{(1)}$ and $K_{\mathrm{II}}^{(1)}$ can be obtained as follows:

$$\begin{cases} K_{\mathrm{I}}^{(1)}=\dfrac{1}{2\beta}\left[G^{(1+2)}\left(K_{\mathrm{I}}^{(2)}=1,K_{\mathrm{II}}^{(2)}=0\right)-G^{(1)}-G^{(2)}\left(K_{\mathrm{I}}^{(2)}=1,K_{\mathrm{II}}^{(2)}=0\right)\right] \\ K_{\mathrm{II}}^{(1)}=\dfrac{1}{2\beta}\left[G^{(1+2)}\left(K_{\mathrm{I}}^{(2)}=0,K_{\mathrm{II}}^{(2)}=1\right)-G^{(1)}-G^{(2)}\left(K_{\mathrm{I}}^{(2)}=0,K_{\mathrm{II}}^{(2)}=1\right)\right] \end{cases} \tag{13}$$

First, K_{I} and K_{II} were calculated by setting $l_k=2a$, where a is the crack length, then those values at $l'=10\mu\mathrm{m}$ were evaluated.

3. MODEL FOR ANALYSIS

Figure 1 shows the loading device for mixed-mode loading proposed by Richard and Benitz[32]. This device enables us to carry out the experiments under seven kinds of mixed-mode loading. Figure 2 shows the three-dimensional model of compact normal and shear specimen. The materials 1 and 2 denote the upper and lower parts of specimen, respectively. Table 1 shows the dimensions of CNS specimens. The plane shape of all specimens is similar. The thickness (B) was 3, 5, 10 or 18 mm for the 60W specimen. The thickness was 5 mm for the 30W, 20W and 15W specimens, and 3 or 5 mm for the 12W specimen.

The three-dimensional finite element analysis was carried out on a half portion of the specimen due to the symmetry with respect to the midsection of the specimen. The layer division ratio was 1:2:3, and the numbers of elements and nodes were 1440, and 7801, respectively. Material 2 was an epoxy resin with Young's modulus E_2 = 2.94GPa, and E_1/E_2 was changed with 1, 3, 6.7, 23.3, 40, 70, 100 and 150. Moreover, the effect of absolute value of Young's modulus and Poisson's ratio were investigated. Poisson's ratios of material 1 and 2 were taken as 0.3, 0.34 and 0.38, respectively. The same loading condition as Kishimoto et al.[36] was employed for the mixed-mode loading. The forces and moments meet the following equilibrium equations.

$$\begin{aligned} P\sin\alpha &= P_2 \\ P\cos\alpha &= P_1 + P_3 \\ P_1 l_1 + P_2 l_2 &= P_3 l_1 \end{aligned} \tag{14}$$

where P_1, P_2 and P_3 were applied to the centre of each bolt hole in order to make the uniform loading. Moreover, the crack front was constrained in the x_1 and x_2 directions. P is the load applied to the bolt hole on the periphery of the loading device shown in Fig.2, and α is a load application angle.

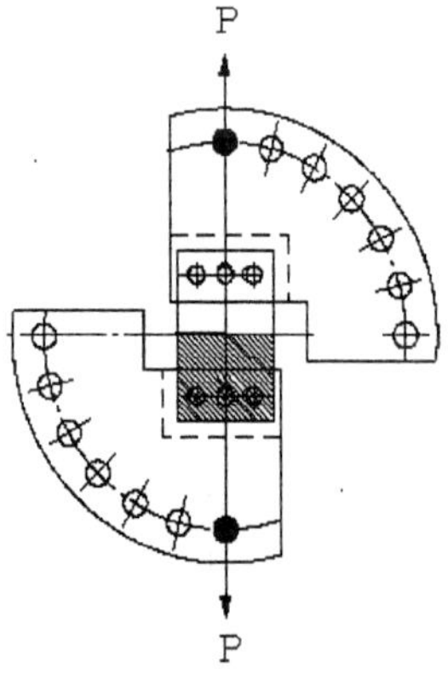

Fig.1 Loading device.

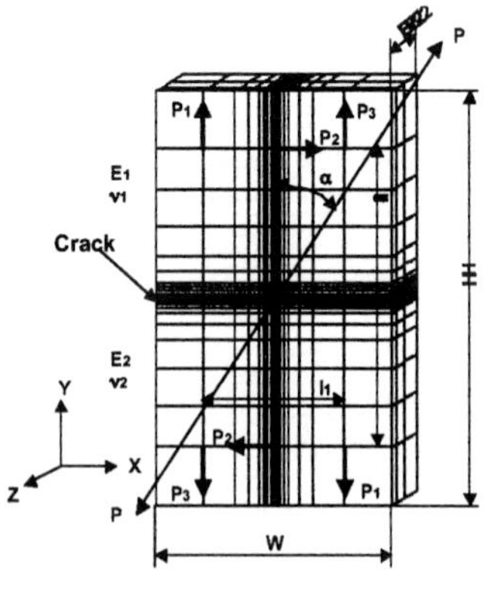

Fig.2 Three-dimensional FEM model.

Table 1 Dimensions of CNS specimens. (in mm)

Specimen	W	H	l_1	l_2
60W	60	100	36	72
30W	30	50	18	36
20W	20	33.3	12	24
15W	15	25	9	18
12W	12	20	7.2	14.4

4. DISCUSSION AND RESULTS

To simplify the comparison between the stress-intensity factors, the normalized stress-intensity factors (F_{I}, F_{II}) were expressed in terms of the stress-intensity factors (K_{I}, K_{II}):

$$F_j = \frac{WB\,K_j}{P\sqrt{\pi a}}, \quad j = \mathrm{I} \text{ and } \mathrm{II} \tag{15}$$

4.1 Distribution of stress-intensity factors along the crack front

Figure 3 shows the distribution of stress-intensity factors along the crack front for ν_1=0.3, ν_2=0.38. Here, Z is the distance from the midsection of the specimen. For E_1/E_2 = 2 (Fig.3(a)), the value of F_{I} is nearly uniform from the midsection of the specimen to $2Z/B$ = 0.4, and decreases gradually toward the free surface. F_{II} is nearly uniform from the midsection to the free surface. Irrespective of specimen dimension, no large differences were in either F_{I} or F_{II}. In the case of E_1/E_2 = 70 (Fig.3(b)), although the difference according to a specimen size in F_{I} and F_{II} is seen, the difference in F_{II} is larger than that in F_{I}. This tendency is almost the same in other

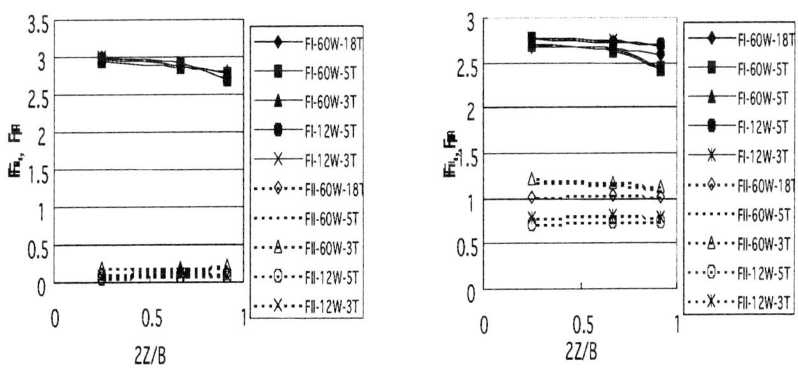

(a) $E_1/E_2 = 2$, $\alpha = 0$ deg

(b) $E_1/E_2 = 70$, $\alpha = 0$ deg

Fig.3 Distribution of F_{I}, F_{II} along the crack front ($\nu_1 = 0.3$, $\nu_2 = 0.38$).

cases and the difference in F_{II} increases gradually with an increase in mixed-mode ratio.

4.2 Polynomial fit of stress-intensity factors

The following things are checked from the previous paper [31].

(i) F_I and F_{II} can be approximated by a cubic expression in α (radian).

(ii) F_I and F_{II} can be approximated by a polynomial in $\sqrt{E_2/E_1}$.

(iii) Both F_I and F_{II} are proportional to the Poisson's ratio.

As mentioned above, in consideration of the dependency of F_I and F_{II} on a, $\sqrt{E_2/E_1}$ and Poisson's ratio, the polynomial approximation by the following equations was performed from the analysis result of 12W-5T.

$$F_{poly} = A_3\alpha^3 + A_2\alpha^2 + A_1\alpha + A_0 \tag{16}$$

$$\begin{aligned} A_3 &= B_{33}x^3 + B_{32}x^2 + B_{31}x + B_{30} \\ A_2 &= B_{23}x^3 + B_{22}x^2 + B_{21}x + B_{20} \\ A_1 &= B_{13}x^3 + B_{12}x^2 + B_{11}x + B_{10} \\ A_0 &= B_{04}x^4 + B_{03}x^3 + B_{02}x^2 + B_{01}x + B_{00} \end{aligned} \tag{17}$$

$$B_{ij} = C_{3ij}\nu_i\nu_j + C_{2ij}\nu_1 + C_{1ij}\nu_2 + C_{0ij} \tag{18}$$

where F_{poly} denotes the stress intensity factor of the polynomial approximation, $x = \sqrt{E_2/E_1}$ and the unit of α is a radian.

The coefficients C_0 ,..., C_3 for F_I and F_{II} are shown in Tables 2 and 3. The stress-intensity factor of CNS specimen of arbitrary Young's modulus, Poisson's ratio, and a load application angle can be evaluated from formulas (18), (17), and (16) using these coefficients. When the specimen form is fixed, the stress-intensity factor can be evaluated by this polynomial approximation with very high accuracy. The difference between the absolute value of the stress-intensity factor obtained from the virtual crack extension method and polynomial approximation was investigated in detail. The numbers of points, which had 0.01 or more differences in F_I and F_{II}, are 64 and 30 in 567 points, respectively. The maximum errors in F_I and F_{II} are 0.024 and 0.02, respectively; neither are large. Thus, the stress-intensity factor of CNS specimen with an interface crack can be simply evaluated by this polynomial approximation with high accuracy.

4.3 Variation of the stress-intensity factors by specimen size

Figure 4 shows the variation of F_I and F_{II} with E_1/E_2 for $\nu_1 = 0.3$ and $\nu_2 = 0.38$. The 3rd numerical value in legend shows a load application angle (α). In a small E_1/E_2 region, F_I decreases rapidly, while F_{II} increases. As E_1/E_2 exceeds 40, both values become nearly constant. For $\alpha = 0$ degree, F_I is not much influenced by the specimen size. However, F_I decreases gradually with an increase in width and load application angle, the difference becomes large. On the other hand, the difference in F_{II} is minimum at $\alpha = 90$ degrees, becomes large with an increase in width. Thus, F_I and F_{II} are influence in reverse by the specimen size.

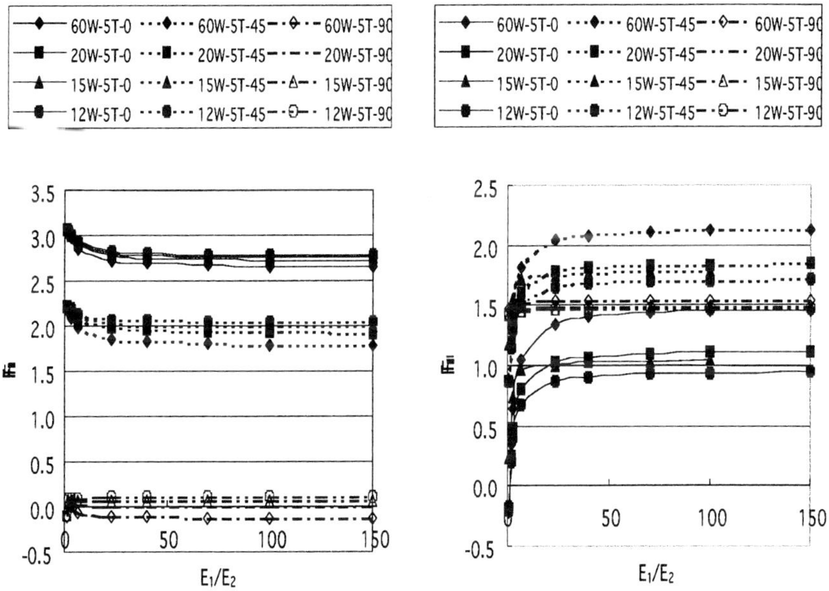

(a) $F_{\rm I}$ (b) $F_{\rm II}$

Fig. 4 Variation of $F_{\rm I}$, $F_{\rm II}$ with E_1/E_2
(ν_1=0.3, ν_2=0.38).

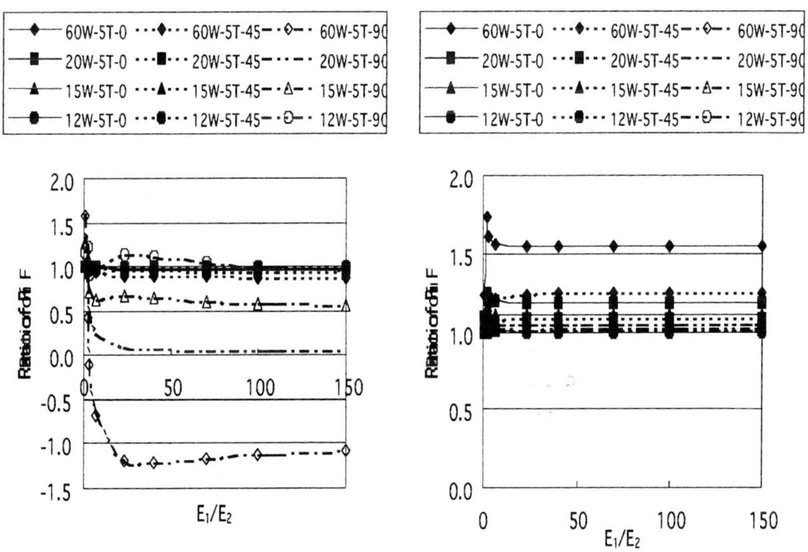

(a) $F_{\rm I}$ (b) $F_{\rm II}$

Fig. 5 Variation of ratio of S.I.F. (VCEM) to
polynomial fit with E_1/E_2.

Figure 5 shows the variation of the ratio of the stress-intensity factors at the central part of each specimen to the stress-intensity factor obtained by the polynomial approximation with E_1/E_2. As this polynomial approximation was obtained about a 12W-5T specimen, error increases with an increase in width. Both F_I values of 12W specimen are in good agreement, with less than1% error. Although the error of 60W specimen increases with an increase in the ratio of Young's modulus, both F_I values are in good agreement with 5% or less error at α = 0 degree, except for a small region of E_1/E_2. At α = 45 degree, the error of the 60W specimen increases. Although the percent error is quite large, the absolute value is small, with 0.13 or less at α = 90 degrees. In F_{II} of the 12W specimen, the error is 1% or less irrespective of the ratio of Young's modulus and load application angle, and both F_{II} values are in good agreement. At α = 0, the maximum error of the 60W specimen is 55% and is quite large. The error decreases with an increase in load application angle. Although the error increases with an increase in thickness at the pure mode II, both F_{II} values of the 60W specimen are in good agreement with an error of 5% or less. The difference between stress intensity factors obtained by the virtual crack extension method and the polynomial fit depends on the specimen thickness. However, the influence was not so large as width.

4.4 Polynomial approximation with consideration to size dependency

The stress intensity factor of the specimen of a fixed configuration can be evaluated with very high accuracy by the polynomial approximation. However, it varies remarkably depending on a specimen size. Therefore, the polynomial approximation was discussed in consideration of the specimen size.

Table 2 Coefficients of polynomials for evaluation of F_I value.

	C_0	C_1	C_2	C_3
B_{00}	1.7869	2.6672	0.5385	-1.7254
B_{01}	-0.2160	0.3707	0.1909	-0.4341
B_{02}	6.9608	-13.368	-7.3775	22.829
B_{03}	-8.9549	15.017	10.698	-28.211
B_{04}	3.0828	-4.1086	-3.5650	7.8195
B_{10}	-1.3817	4.5823	2.0173	-6.5599
B_{11}	5.9783	-19.769	-15.455	50.955
B_{12}	-16.422	57.525	44.351	-155.15
B_{13}	15.251	-52.539	-40.594	140.00
B_{20}	0.0530	-5.2821	-3.0900	10.262
B_{21}	-14.102	46.192	36.061	-120.35
B_{22}	42.848	-144.41	-114.52	382.52
B_{23}	-38.234	129.26	101.76	-341.01
B_{30}	-0.2607	1.8687	1.4485	-4.8647
B_{31}	7.6055	-25.274	-19.707	66.084
B_{32}	-24.666	82.370	64.674	-215.75
B_{33}	21.945	-73.444	-57.533	192.31

Table 3 Coefficients of polynomials for evaluation of F_{II} value.

	C_0	C_1	C_2	C_3
B_{00}	3.1574	-6.0566	-1.4197	4.6053
B_{01}	1.0603	-2.3721	-2.2108	5.4948
B_{02}	-12.569	20.135	21.975	-40.698
B_{03}	12.886	-23.726	-26.756	54.690
B_{04}	-4.7848	10.336	11.516	-26.068
B_{10}	1.6781	-0.3672	-0.1350	0.3608
B_{11}	-0.2126	0.2332	0.3456	-0.2382
B_{12}	-0.5496	0.7491	0.3008	-0.1000
B_{13}	0.4358	-0.5385	-0.2781	-0.0673
B_{20}	-2.0743	3.8051	1.0316	-3.1686
B_{21}	0.3788	-0.7418	-1.2005	2.3311
B_{22}	3.4614	-3.2654	-3.0258	1.5386
B_{23}	-1.7073	1.2120	1.3153	0.4921
B_{30}	0.1967	-0.6955	-0.1880	0.5782
B_{31}	-0.0601	0.1363	0.2445	-0.5182
B_{32}	-0.5639	0.4683	0.3861	0.0172
B_{33}	0.2478	-0.1142	-0.1001	-0.3105

Since the stress-intensity factor was influenced more by width than by thickness, the following polynomial fit was conducted

$$\frac{F}{F_{poly}} = D_0 + D_1 W + D_2 W\alpha + D_3 W\sqrt{\frac{E_2}{E_1}} + D_4 W\alpha \frac{E_2}{E_1} \tag{19}$$

The coefficients D_0, ..., D_4 are shown in Table 4. Tables 5 and 6 show the difference of the absolute value of stress-intensity factors obtained by the virtual crack extension method and eq. (19) for ν_1=0.3 and ν_2=0.38. For 60W-18T and 60W-5T and $\alpha = 0$. The error arises in the 2nd decimal place at the other parts. For F_{II}, many points with 0.1 or more error are observed in 60W-18T. However, the error is not so large as 0.2. There are 37 points with 0.1 or more error among 378 points; this is larger than for F_{I}.

Table 4 Coefficients of polynomials for evaluation of F_{I} and F_{II}.

	F_{I}	F_{II}
D_0	1.006	9.952×10^{-1}
D_1	-1.261×10^{-4}	7.985×10^{-3}
D_2	-2.537×10^{-3}	-4.884×10^{-3}
D_3	-1.518×10^{-4}	-5.972×10^{-3}
D_4	3.368×10^{-3}	3.229×10^{-3}

Table 5 Differences of F_{I} between VCEM and Polynomial fit (ν_1=0.3, ν_2=0.38).

(1) 60W-18T

		E_1/E_2								
		1	2	3	6.667	23.33	40	70	100	150
α (deg)	0	0.028	0.024	0.001	-0.048	-0.100	-0.110	-0.117	-0.120	-0.122
	15	0.014	0.049	0.041	0.006	-0.041	-0.052	-0.059	-0.062	-0.065
	30	-0.016	0.077	0.083	0.061	0.020	0.010	0.003	-0.001	-0.004
	45	-0.046	0.088	0.101	0.086	0.051	0.042	0.035	0.031	0.028
	60	-0.048	0.072	0.080	0.066	0.036	0.027	0.021	0.017	0.014
	75	0.015	0.035	0.022	-0.002	-0.028	-0.036	-0.043	-0.047	-0.051
	90	0.043	-0.011	-0.065	-0.052	-0.011	-0.008	-0.009	-0.012	-0.015

(2) 60W-5T

		E_1/E_2								
		1	2	3	6.667	23.33	40	70	100	150
α (deg)	0	0.003	0.003	-0.015	-0.056	-0.099	-0.108	-0.113	-0.116	-0.118
	15	-0.013	0.023	0.017	-0.012	-0.053	-0.062	-0.069	-0.071	-0.074
	30	-0.043	0.048	0.054	0.034	-0.003	-0.013	-0.020	-0.023	-0.025
	45	-0.071	0.058	0.069	0.052	0.017	0.008	0.001	-0.002	-0.005
	60	-0.069	0.043	0.047	0.027	-0.006	-0.015	-0.022	-0.025	-0.028
	75	0.000	0.008	-0.011	-0.042	-0.074	-0.083	-0.091	-0.096	-0.100
	90	0.035	-0.033	-0.073	-0.013	0.038	0.042	0.042	0.039	0.036

Table 5 Cont.

(3) 30W-5T

		E_1/E_2								
		1	2	3	6.667	23.33	40	70	100	150
α (deg)	0	0.008	0.010	-0.003	-0.031	-0.061	-0.067	-0.071	-0.072	-0.074
	15	0.008	0.015	0.006	-0.017	-0.046	-0.053	-0.057	-0.059	-0.061
	30	-0.010	0.029	0.025	0.006	-0.022	-0.029	-0.034	-0.036	-0.039
	45	-0.033	0.033	0.032	0.013	-0.013	-0.019	-0.025	-0.027	-0.029
	60	-0.035	0.022	0.016	-0.004	-0.026	-0.033	-0.038	-0.041	-0.044
	75	0.017	0.001	-0.019	-0.043	-0.061	-0.068	-0.074	-0.078	-0.082
	90	0.025	-0.015	-0.062	-0.087	-0.057	-0.058	-0.062	-0.066	-0.070

(4) 20W-5T

		E_1/E_2								
		1	2	3	6.667	23.33	40	70	100	150
α (deg)	0	0.003	0.006	-0.004	-0.028	-0.054	-0.060	-0.064	-0.065	-0.066
	15	0.008	0.007	-0.001	-0.022	-0.048	-0.054	-0.059	-0.061	-0.063
	30	-0.006	0.019	0.013	-0.005	-0.031	-0.037	-0.042	-0.044	-0.046
	45	-0.026	0.023	0.019	0.001	-0.023	-0.029	-0.034	-0.036	-0.038
	60	-0.029	0.015	0.007	-0.011	-0.031	-0.037	-0.042	-0.045	-0.048
	75	0.019	0.001	-0.018	-0.038	-0.053	-0.058	-0.065	-0.068	-0.072
	90	0.021	-0.005	-0.046	-0.069	-0.076	-0.082	-0.090	-0.096	-0.102

(5) 15W-5T

		E_1/E_2								
		1	2	3	6.667	23.33	40	70	100	150
					0	-0.027	-0.030	-0.031	-0.032	-0.032
	15	0.003	0.002	-0.002	-0.008	-0.017	-0.020	-0.022	-0.023	-0.024
	30	-0.008	0.014	0.014	0.010	0.003	0.001	-0.001	-0.003	-0.004
	45	-0.027	0.020	0.022	0.019	0.014	0.012	0.009	0.008	0.007
	60	-0.029	0.015	0.014	0.011	0.009	0.007	0.004	0.002	0.000
	75	0.018	0.006	-0.005	-0.009	-0.008	-0.011	-0.015	-0.017	-0.021
	90	0.017	0.007	-0.025	-0.032	-0.025	-0.028	-0.035	-0.040	-0.045

(6) 12W-5T

		E_1/E_2								
		1	2	3	6.667	23.33	40	70	100	150
					0	-0.009	-0.009	-0.009	-0.009	-0.009
	15	-0.005	-0.005	-0.004	0.003	0.006	0.007	0.006	0.006	0.006
	30	-0.015	0.009	0.014	0.024	0.031	0.031	0.031	0.031	0.030
	45	-0.034	0.017	0.025	0.037	0.045	0.046	0.046	0.045	0.044
	60	-0.036	0.016	0.021	0.033	0.043	0.044	0.043	0.042	0.041
	75	0.011	0.011	0.007	0.016	0.030	0.030	0.028	0.025	0.022
	90	0.009	0.016	-0.007	-0.001	0.017	0.015	0.010	0.006	0.000

Table 6 Differences of F_{II} between VCEM and Polynomial fit (ν_1=0.3, ν_2=0.38).

(1) 60W-18T

		E_1/E_2								
		1	2	3	6.667	23.33	40	70	100	150
α (deg)	0	0.059	0.012	0.003	-0.038	-0.089	-0.101	-0.108	-0.111	-0.113
	15	-0.095	-0.070	-0.087	-0.127	-0.174	-0.185	-0.192	-0.194	-0.196
	30	-0.102	-0.104	-0.121	-0.153	-0.189	-0.197	-0.202	-0.204	-0.206
	45	-0.084	-0.103	-0.117	-0.138	-0.161	-0.166	-0.170	-0.171	-0.172
	60	-0.048	-0.076	-0.086	-0.099	-0.112	-0.115	-0.117	-0.118	-0.118
	75	0.001	-0.031	-0.039	-0.049	-0.059	-0.060	-0.062	-0.062	-0.062
	90	0.068	0.028	0.018	0.003	-0.009	-0.010	-0.011	-0.011	-0.011

(2) 60W-5T

		E_1/E_2								
		1	2	3	6.667	23.33	40	70	100	150
α (deg)	0	0.022	0.087	0.100	0.096	0.080	0.075	0.073	0.072	0.072
	15	-0.047	0.016	0.021	0.017	0.005	0.001	-0.001	-0.001	-0.001
	30	-0.046	-0.013	-0.010	-0.008	-0.012	-0.014	-0.014	-0.015	-0.014
	45	-0.024	-0.013	-0.010	-0.002	0.002	0.002	0.002	0.002	0.003
	60	0.012	0.007	0.009	0.018	0.025	0.026	0.027	0.027	0.028
	75	0.057	0.039	0.038	0.041	0.044	0.045	0.045	0.046	0.046
	90	0.112	0.080	0.072	0.060	0.052	0.051	0.051	0.051	0.051

(3) 30W-5T

		E_1/E_2								
		1	2	3	6.667	23.33	40	70	100	150
α (deg)	0	0.015	0.063	0.079	0.090	0.088	0.087	0.086	0.086	0.086
	15	-0.028	0.026	0.037	0.047	0.047	0.046	0.045	0.045	0.045
	30	-0.026	0.011	0.020	0.032	0.036	0.036	0.036	0.036	0.036
	45	-0.013	0.009	0.018	0.031	0.037	0.038	0.038	0.038	0.039
	60	0.006	0.015	0.021	0.031	0.038	0.039	0.039	0.040	0.040
	75	0.028	0.024	0.027	0.031	0.034	0.034	0.034	0.035	0.035
	90	0.058	0.042	0.039	0.033	0.028	0.027	0.027	0.028	0.028

(4) 20W-5T

		E_1/E_2								
		1	2	3	6.667	23.33	40	70	100	150
α (deg)	0	0.011	0.030	0.035	0.035	0.028	0.026	0.025	0.024	0.024
	15	-0.022	0.003	0.005	0.005	0.000	-0.002	-0.004	-0.004	-0.004
	30	-0.020	-0.005	-0.004	-0.001	-0.003	-0.005	-0.006	-0.006	-0.006
	45	-0.011	-0.004	-0.003	0.002	0.003	0.002	0.002	0.002	0.002
	60	0.000	0.000	0.002	0.007	0.008	0.008	0.008	0.008	0.008
	75	0.014	0.008	0.009	0.011	0.011	0.011	0.011	0.012	0.012
	90	0.035	0.024	0.023	0.021	0.017	0.017	0.017	0.017	0.018

Table 6 Cont.

(5) 15W-5T

		E_1/E_2								
		1	2	3	6.667	23.33	40	70	100	150
α (deg)	0	0.006	0.015	0.013	0.005	-0.007	-0.01	-0.012	-0.012	-0.012
	15	-0.015	-0.006	-0.011	-0.019	-0.029	-0.032	-0.034	-0.034	-0.035
	30	-0.013	-0.011	-0.016	-0.021	-0.028	-0.031	-0.032	-0.032	-0.032
	45	-0.007	-0.009	-0.012	-0.015	-0.019	-0.021	-0.022	-0.022	-0.022
	60	0.001	-0.004	-0.007	-0.008	-0.01	-0.011	-0.012	-0.012	-0.012
	75	0.011	0.003	0.001	-0.001	-0.003	-0.003	-0.003	-0.003	-0.003
	90	0.027	0.017	0.016	0.013	0.010	0.01	0.01	0.01	0.011

(6) 12W-5T

		E_1/E_2								
		1	2	3	6.667	23.33	40	70	100	150
α (deg)	0	-0.006	-0.009	-0.028	-0.054	-0.078	-0.083	-0.085	-0.086	-0.087
	15	-0.003	-0.027	-0.048	-0.073	-0.096	-0.100	-0.103	-0.104	-0.105
	30	-0.003	-0.030	-0.048	-0.069	-0.089	-0.092	-0.094	-0.095	-0.096
	45	0.000	-0.025	-0.040	-0.056	-0.070	-0.073	-0.074	-0.075	-0.075
	60	0.003	-0.018	-0.029	-0.039	-0.049	-0.051	-0.052	-0.052	-0.052
	75	0.008	-0.008	-0.014	-0.021	-0.027	-0.028	-0.028	-0.028	-0.028
	90	0.018	0.010	0.009	0.006	0.003	0.003	0.003	0.004	0.004

Although several points with 0.1 or more error are seen in 12W-5T, the error is 8% or less. Even if the plane dimension of the specimen becomes large proportionally in order of 12W, 15W, 20W and 30W, the stress-intensity factor can be evaluated by this polynomial approximation with 0.1 or less error. Thus, this polynomial approximation is considered to be very effective for the evaluation of the stress-intensity factor in the fracture-toughness test by the CNS specimen with an interface crack.

5. CONCLUSIONS

1. The stress intensity factor of the specimen of a fixed configuration can be evaluated with very high accuracy by a polynomial approximation.

2. The stress-intensity factor is greatly influenced by the plane size rather than by thickness.

3. The stress-intensity factor of CNS specimen can be evaluated with a sufficient accuracy by this polynomial approximation. Therefore, it is considered that this polynomial approximation is effective for the simple evaluation of stress-intensity factor in a fracture toughness test.

References

(1) Williams, M.L., The Stresses around a Fault or Crack in Dissimilar Media, Bull. Seismol. Soc. America, Vol.49(1959), p.199-204.

(2) Erdogan, F.J., Stress Distribution in a Homogeneous Elastic Plane With Cracks, J. Appl. Mech., Vol.30(1963), p.232-236.

(3) Erdogan, F.J., Stress Distribution in Bonded Dissimilar Materials with Cracks, J. Appl. Mech., Vol.32(1965), p.403 410.

(4) Rice, J.R. and Sih, G.C., Plane Problems of Cracks in Dissimilar Media, J. Appt. Mech., Vol.32(1965), p.418-423.

(5) Yau, J.F., Wang, S.S. and Corten, H.T., A Mixed-Mode Crack Analysis of Isotropic Solids Using Conservation Laws of Elasticity, J. Appt. Mech., Vol.47(1980), p.335-341.

(6) Yau, J.F. and Wang, S.S., An Analysis of Interface Cracks Between Dissimilar Isotropic Materials Using Conservation Integrals in Elasticity, Eng. Fract. Mech., Vol.20, No.3(1984), p.423-432.

(7) Sun. C.T. and Jih, C.J., On Strain Energy Release Rates for Interfacial Cracks in Bi-Material Media, Eng. Fract. Mech., Vol.28, No.l(1987), p.13-20.

(8) Hutchinson, J.W., Mear, M.E. and Rice, J.R., Crack Paralleling an Interface Between Dissimilar Materials, J. Appt. Mech., Vol.54(1987), p. 828-832.

(9) Gautesen, A.K. and Dundurs, J., The Interface Crack in a Tension Field, J. Appl. Mech., Vol.54(1987), p.93-98.

(10) Rice, J.R., Elastic Fracture Mechanics Concepts for Interfacial Cracks, J. Appl. Mech., Vol.55 (1988), p.98-103.

(11) Sih, CF. and Asaro, R.j., Elastic-Plastic Analysis of Cracks on Bimaterial Interfaces: Part I-Small Scale Yielding, J. Appl. Mech., Vol.55(1988), p.299-316.

(12) Delale, F. and Erdogan, F., On the Mechanical Modeling of the Interfacial Region in Bonded Half-Planes, J. Appt. Mech., Vol.55(1988), p.317-324.

(13) Charalambides, P.G., Lund, J., Evans, A.G. and McMeeking, R.M., A Test Specimen for Determining the Fracture Resistance of Bimaterial Interfaces, J. Appl. Mech., Vol.56(1989), p.77 82.

(14) Matos, P.P.L., Mch4eeking, R.M., Charalambides, P.G. and Dropy, M.D., A Method for Calculating Stress Intensities in Bimaterial Fracture, Int. J. Fract., Vol.40(1989), p.235-254.

(15) Yuuki, R. and Cho, S.B., Boundary Element Analysis of Stress Intensity Factors for an Interface Crack in Dissimilar Materials, Trans. Jpn. Soc. Mech. Eng., (in Japanese), Vol. 55, No.510, A(1989), p.340-347.

(16) Yuuki, R. and Xu, J.Q., Logarithmic Singularity on Thermal or Residual Stress at the Interface Edge Point of Bonded Dissimilar Materials, Trans. Jpn. Soc. Mech. Eng., (in Japanese), Vol.58, No.556, A(1992), p.2394-2400.

(17) Yuuki, R., Liu, J.Q., Xu, J.Q. and Kim, Y.H., Study on the Mixed Mode Fracture Criterion of Interface Crack, J. Soc. Mat. Sci., (in Japanese), Vol.43, No.493(1994), p.1206-1210.

(18) Chen, D.H. and Mori, Y., Stress Singularities for a V-Notch with Its Tip on the Bimaterial Interface, Trans. Jpn. Soc. Mech. Eng., (in Japanese), Vol.58, No.556, A(1992), p.2381-2386.

(19) Chen, D.H. and Nishitani, H., Stress Intensity Factors of a Crack Meeting the Bimaterial Interface, Trans. Jpn. Soc. Mech. Eng., (in Japanese), Vol.59, No.558, A(1993), p.325 331.

(20) Chen, D.H. and Mori, Y., Stress Singularities for Crack with Tip on Bimaterial Interface of Isotropic and Anisotropic Phases, Trans. Jpn. Soc. Mech. Eng., (in Japanese), Vol.60, No.578, A(1994), p.2228 2235.

(21) Aoki, S., Kishimoto, K. and Takeuchi, N., An Elastic-Plastic Finite Element Analysis of a Crack on a Bimaterial Interface, Trans. Jpn. Soc. Mech. Eng., (in Japanese), Vol.56, No.523, A(1990), p.638-645.

(22) Aoki, S. and Ishii, N., Interface Crack under Mixed Mode Loading, Trans. Jpn. Soc. Mech. Eng., (in Japanese), Vol.58, No.555, A(1992), p.2056-2061.

(23) Miyazaki, N., Ikeda, T., Soda, T. and Munakata, T., Stress Intensity Factor Analysis of Interface Crack Using Boundary Element Method (1st Report, Application of Virtual Crack Extension Method), Trans. Jpn. Soc. Mech. Eng., (in Japanese), Vol.57, No.541, A(1991), p.2063-2069.

(24) Miyazaki, N., Ikeda, T., Soda, T. and Munakata, T., Stress Intensity Factor Analysis of Interface Crack Using Boundary Element Method (2nd Report, Application of Contour-Integral Method, Trans. Jpn. Soc. Mech. Eng., (in Japanese), Vol.57, No.544, A(1991), p.2903-2910.

(25) Ikeda, T., Miyazaki, N,, Soda, T. and Munakata, T., Mixed Mode Fracture Criteria of Interface Crack between Dissimilar Materials, Trans. Jpn. Soc. Mech. Eng., (in Japanese), Vol.58, No.555, A(1992), p.2080-2087.

(26) Ikeda, T., Miyazaki, N., Yamashita, A. and Munakata, T., Stress Intensity Factor Analysis of an Interface Crack in an Adhesive Joint by Combination of the Boundary Element and Finite Element Methods, Trans. Jpn. Soc. Mech. Eng., (in Japanese), Vol.60, No.578, A(1994), p.2220-2227.

(27) Ikeda, T., Komohara, Y. and Miyazaki, N., Stress Intensity Factor Analysis of an Interface Crack between Dissimilar Materials under Thermal Stress Condition by Virtual Crack Extension Method, Trans. Jpn. Soc. Mech. Eng., (in Japanese), Vol.63 No.611(1997, p.1377-1384.

(28) Noda, N., Oda, K., Higuchi, T. and Tanaka, A., Analysis of Stress Intensity Factors of Interface Cracks and Angular Corners Using the Singular Integral Equation of the Body Force Method, Trans. Jpn. Soc. Mech. Eng., (in Japanese), Vol.60, No.578, A(1994), p.2213-2219.

(29) Machida, K., Stress Intensity Factors of a Mixed-Mode Interface Crack by a Finite Element Analysis, Trans. Jpn. Soc. Mech. Eng., (in Japanese), Vol.62, No.599, A(1996), p.1565-1570.

(30) Machida, K., Stress Intensity Factors of 3-D Interface Crack under Mixed-Mode Loading, Trans. Jpn. Soc. Mech. Eng., (in Japanese), Vol.63, No.606, A(1997), p.302-307.

(31) Machida, K., Polynomial Evaluation of Stress Intensity Factors of CNS Specimen with an Interface Crack Subject to Mixed-Mode Loading, Trans. Jpn. Soc. Mech. Eng., (in Japanese), Vol.64, No.618, A(1998), p.290-296.

(32) Richard, H.A. and Benitz, K., A Loading Device for the Creation of Mixed Mode in Fracture Mechanics, Int. J. Fract., Vol.22(1983), p.R55-R58.

(33) Parks, D.M., A Stiffness Derivative Finite Element Technique for Determination of Elastic Crack Tip Stress Intensity Factors, Int. J. Fract,, Vol.10(1974), p.487-502.

(34) Rybicki, E.F. and Kanninen, M.F., A Finite Element Calculation of Stress Intensity Factors by a Modified Crack Closure Integral, Eng. Fract. Mech., Vol.9 (1977), p.931-938.

(35) Shivakumar, K.N., Tan, P.W. and Newman, J,C. Jr., A Virtual Crack Closure Technique for Calculating Stress Intensity Factors for Cracked Three Dimensional Bodies, Int. J. Fract., Vol.36(1988), p.R43-R50.

(36) Kishimoto, K., Fukano, H., Yoshida, T. and Aoki, S., Elastic-Plastic Fracture Behavior of an Aluminum Alloy under Mixed Mode Conditions, Trans. Jpn. Soc. Mech. Eng., (in Japanese), Vol.56, No.524 (1990), p.957-965.

PENNY–SHAPED EQUILIBRIUM CRACKS IN ALIGNED COMPOSITES

M. A. Grekov, N. F. Morozov and N. V. Ponikarov
St.Petersburg State University, Russia

Abstract. A model is suggested for a partially bridged penny-shaped crack (axisymmetric problem) in a brittle aligned material like for a composite ceramic. Two different fracture criteria for its components (matrix and fiber) are accepted. On the basis of an analytical solution for a homogeneous anisotropic body, a force-separation law, and Novozhilov's brittle fracture criterion, the variation intervals of a diameter of an equilibrium crack and the width of a bridged crack part are estimated. It is shown that, like a fracture toughness, the critical width of the bridged crack part can be accepted as a constant parameter for a composite material reinforced by fibers. The value of this parameter for a penny-shaped crack is the same as for a crack under plane deformation. For two types of ceramics the variation intervals of a bridged part of a critical crack are found, and a dependence of an ultimate load upon the size of the crack in 2-D and axisymmetric problems is presented.

1. INTRODUCTION

The toughness of brittle solids such as ceramics can be considerably increased by use of fibers which bridge a crack. Experiments show that a crack normal to the fibers spreads in its plane under loading symmetrical to this plane and can be partially or fully bridged, but consideration of a partially bridged crack in a brittle solid is more interesting for applications since such a crack is mostly the forerunner of a failure with an unstable crack growth.

Using Novozhilov's discrete model of a continuum (Novozhilov, 1969a, b), Grekov and Morozov (2000) considered partially bridged penny-shaped cracks in a homogeneous isotropic body, and estimated variation interval of a crack diameter under fixed load for equilibrium cracks. Novozhilov's approach correlates closely with Thomson's theory of lattice trapping, and the variation interval of applied load under which the crack is in equilibrium (Thomson at al., 1971), Morozov and Paukshto, 1988), and let us solve some applications (see Morozov and Paukshto, 1991, Morozov et al., 1997). Novozhilov–Thomson's model is based on an account of interacting forces between two rows of atoms near the edge of a microcrack. The extension of this model to a crack in a homogeneous body leads to a consideration of interacting forces between the crack surfaces. The nature of these forces does not have a clear physical meaning, so it is not defined precisely in Grekov and Morozov (2000). At the same time, Novozhilov–Thomson's model is ideally suited to composites (like ceramics) that consist of a brittle matrix reinforced by unidirectional fibers. For such

P. Ståhle and K.G. Sundin (eds.), IUTAM Symposium on Field Analyses for Determination of Material Parameters - Experimental and Numerical Aspects, 203–216.

materials interacting forces are natural reaction forces of fibers connecting surfaces of a bridged crack near its edge (Morozov et al., 1997).

The size of a bridged crack part depends on the stress-strain behavior and the strength of a fiber, and on fiber-to-matrix adhesion as well. There are a few solutions obtained for a transversely isotropic elastic body with a bridged crack both under plane (e.g. Nemat-Nasser and Hori, 1987, Morozov et al., 1997) and axisymmetrical (e.g. Movchan and Willis, 1998) deformation. As a rule, authors restrict their consideration to either an analytical or a computional solution of a corresponding boundary problem and (except for Morozov et al., 1997) do not touch upon the subject of fracture and crack propagation. Usually, the force-separation law that characterizes the restraining traction provided by an unbroken fiber, is either postulated (Nemat-Nasser and Hori, 1998, Movchan and Willis, 1998) or found from a solution of the corresponding problem of drawing of a fiber from matrix with allowance for elastic-plastic characteristics of a fiber and fiber-to-matrix adhesion (Budiansky et al., 1995).

The present article continues the work of Grekov and Morozov (2000) on a circular crack in an anisotropic material. As distinguished from that work we suppose now that the aligned material is inhomogeneous. This inhomogeneity, or structural imperfection of the material, is taken into account by means of Novozhilov-Thomson's hybrid method (Novozhilov, 1969a, b, Thomson et al., 1971). Namely, the crack growth is determined by two events: the breakage of extreme fibers in a bridged crack part, and failure of a brittle matrix in a fracture zone adjoining the edge of the crack. We assume also that, before unstable crack growth, the size of the bridged zone becomes much smaller than the radius of the crack. The same approach was used by Morozov et al. (1997) for the case of plane deformation. Note that this assumption is not *sine qua non*, and one can repeat all the present reasoning and obtain analogue results without it.

2. THE STATEMENT OF AN AXISYMMETRIC PROBLEM

Consider an elastic composite (e.g. ceramic composite) consisting of a brittle matrix and brittle unidirectional fibers. The matrix contains a penny-shaped crack $\Gamma = \{(x_1, x_2, x_3):\ x_3 = 0,\ \rho^2 = x_1^2 + x_2^2 \le a^2\}$ (Fig. 1) normal to the axis of the fibers. Generally, the crack is opened by remote loading $\sigma_{33}^{\infty} = p$ and two point opposite forces, P, applied at the center of the crack faces normally to the crack . Most of the fibers that had bonded to the crack surfaces have been broken and only a narrow annulus $\Gamma_0 = \{(x_1, x_2, x_3):\ x_3 = 0,\ b \le \rho \le a\}$ of width $\Delta = a - b$ is bridged by the fibers (Fig. 1). We simualate their resistance to the crack opening by the action of a normal traction, σ_0, uniformly and continuously distributed over Γ_0 (Fig. 2).

The stress in the boundary bridging fibers (for $\rho = b$) is equal to the tensile strength, R_f, of fiber material. At the same time for the crack opening displacement, $2w$, we have

$$w(b) = w_0, \tag{1}$$

where w_0 is a constant.

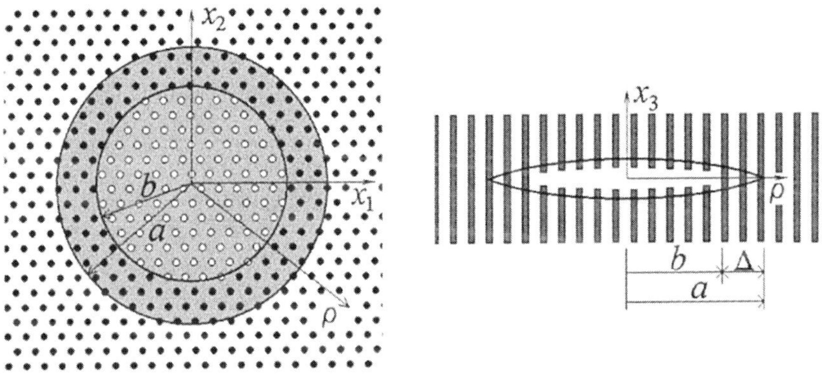

Fig. 1. A penny-shaped bridged crack in composite reinforced by fiber (top view on the left and side view on the right).

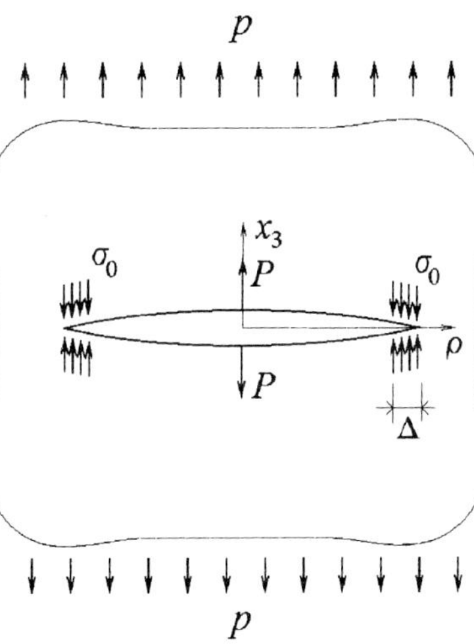

Fig. 2. The model of a partially bridged crack under loading.

If $w(b) < w_0$ the crack is considered as closed. The inequality $w(\rho) > w_0$ mean, that the fibers are broken on the circle of radius ρ. So, $w(\rho) > w_0$ if $\rho < b$, and $w(\rho) < w_0$ if $b < \rho < a$.

The value σ_0 is determined from the force-separation law (see Fig. 3) $\sigma = \sigma(w)$, related to pulling out of a fiber from the matrix, by means of the formula

$$\sigma_0 = \frac{c}{w_0 - w_B} \int_{w_B}^{w_0} \sigma(t)dt . \tag{2}$$

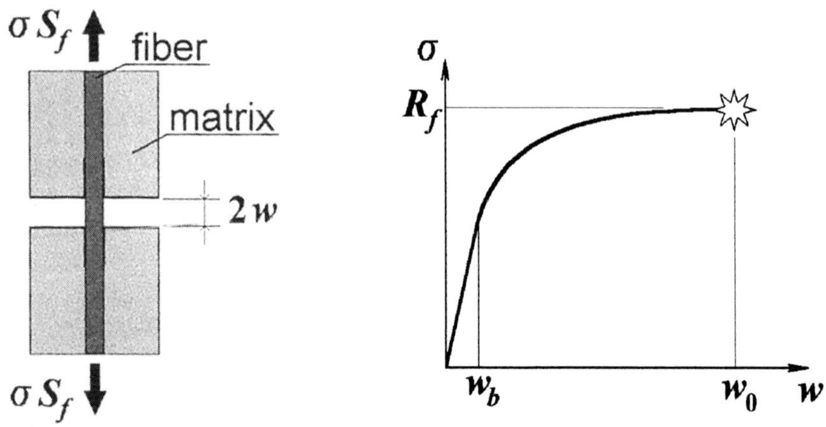

Fig. 3. Pulling out of a fiber from matrix and corresponding force-separation law.

Here $\sigma(w)$ is a tensile stress in a fiber; the displacement w_B of the upper and lower crack faces corresponds to the beginning of a nonlinear part of the relation $\sigma = \sigma(w)$; c is a fiber volume fraction in the composite. The values w_B and w_0 depend on the properties of fiber-to-matrix adhesion and the elastic constants of each component.

The physical meaning of (2) is as follows: when the crack is opened and the edge fibers are pulled out over the interval (w_B, w_0), the work of the traction σ_0 is equal to the work of the real fiber stress σ. Moreover formula (2) takes into account a fiber volume fraction in the composite. Note that the substitution of the proper function $\sigma_0(w)$, which can be obtained only from a solution of an appropriate i ntegral equation (see Nemat-Nasser and Hori (1987), and Movchan and Willis (1998)), by the constant value σ_0 from (2) is similar to Novozhilov's substitution of the descending brunch of the atom interaction law by a step-function (see Novozhilov (1969a, b)).

Following Novozhilov's fracture criterion, we suppose that the fracture of matrix the near the crack tip does not occur if

$$\int_a^{a+D} \sigma_{33}(\rho) d\rho \leq D R_m . \tag{3}$$

Here R_m is the tensile strength of matrix, D is the size of the matrix fracture zone. If condition (3) is violated the crack will grow.

As we have brittle fracture of the matrix, we determine value D by the formula

$$D = kD_m , \qquad D_m \equiv \frac{2}{\pi}\left(\frac{K_{1c}^m}{R_m}\right)^2 . \tag{4}$$

where K_{1c}^{m} is a fracture toughness of matrix and k is a coefficient that reflects the influence of composite structural inhomogeneity upon the fracture processes in the matrix.

The equality $D = D_m$ follows from a formal correlation of Novozhilov's brittle fracture criterion and Irvin's criterion for a homogeneous isotropic material. For fiber-reinforced composite, it means that reinforcement does not affect fracture processes in the matrix.

2. INTERVALS OF EQUILIBRIUM CRACK SIZES

Like Grekov and Morozov (2000) we are guided by Panasyuk (1968) and Fabrikant (1990), and construct our solution for an equilibrium crack being in a homogeneous transversely isotropic half space under the given loading (Fig. 2). For our purposes we need expressions only for crack opening displacement, $2w$, and stress σ_{33}, in the plane $x_3 = 0$; these are as follows:

$$w(\rho) = 2H\left\{2p\sqrt{a^2 - \rho^2} - 2\sigma_0 F(\rho) + \frac{P}{\pi\rho}\operatorname{arctg}\frac{\sqrt{a^2 - \rho^2}}{\rho}\right\}, \tag{5}$$

$$\sigma_{33}(\rho) = \frac{2}{\pi}\left[\frac{ap - \sigma_0\sqrt{a^2 - b^2}}{\sqrt{\rho^2 - a^2}} + \frac{\pi p}{2} - p\arcsin\left(\frac{a}{\rho}\right) + \sigma_0 \arcsin\sqrt{\frac{a^2 - b^2}{\rho^2 - b^2}}\right] +$$

$$+ \frac{Pa}{\pi^2\rho^2\sqrt{\rho^2 - a^2}}, \qquad \rho > a. \tag{6}$$

Here

$$F(\rho) = \int_r^a \sqrt{\frac{t^2 - b^2}{t^2 - \rho^2}}\, dt, \qquad r = \begin{cases} b, & 0 < \rho < b \\ \rho, & b < \rho < a \end{cases}.$$

According to Fabricant (1990), the coefficient H, that characterizes the stiffness of anisotropic materials, is determined by

$$H = \frac{(\gamma_1 + \gamma_2)A_{11}}{2\pi(A_{11}A_{33} - A_{13}^2)}, \tag{7}$$

where $\gamma_k^2 = \dfrac{m_k A_{33}}{m_k A_{44} + A_{13} + A_{44}} = \dfrac{A_{44} + m_k(A_{13} + A_{44})}{A_{11}}$, $k = 1, 2$; A_{jj} are components of elastic compliance tensor along the principal axes of anisotropy.

For materials reinforced by cylindrical parallel long fibers, the coefficients A_{jj} can be obtained from elastic coefficients and volume fractions of composite components (see Christensen (1979)). For ceramics-based composites, the Young´s moduli of the fibers, E_f, and of the matrix usually differ by a factor somewhat less than three; the Poisson's ratios, ν_f, and, ν_m, are equal. So, such material is slightly anisotropic, and we suggest an approximation for H simpler than formula (7), in conjunction with a nontrivial approximate method for the determination of A_{jj}:

$$H = \frac{1 - \nu_f^2}{\pi[cE_f + (1 - c)E_m]} \tag{8}$$

One can show that formulas (7) and (8) give approximately the same calculation errors for H. When $E_f = E_m$ equation (8) is identical to the analogous equation for isotropic material (see Morozov and Grekov (2000)). All computed results presented below were obtained with formula (8).

Now we will find the condition of equilibrium state of the crack in the composite. Consider two cases: remote loading, p ($P = 0$), and the effect of concentrated forces, P ($p = 0$).

The crack is opened by remote load $\sigma_{33}^{\infty} = p$. Substitute (5) into (1), (6) into (3), taking into account that $P = 0$. Then, supposing that $\Delta / a << 1$, $D / a << 1$ we obtain:

$$\frac{2p}{\sigma_0}\left(\frac{d}{D}\right)^{1/2} = g_1^a(\alpha), \tag{9}$$

$$\frac{2p}{\sigma_0}\left(\frac{d}{D}\right)^{1/2} \le g_2^a(\alpha), \tag{10}$$

where d is a crack diameter, $\alpha = \Delta / D$ is the reduced width of the bridged zone.

Functions g_1^a, g_2^a are given by formulas

$$g_1^a(\alpha) = \frac{\beta}{2\sqrt{\alpha}} + 2\sqrt{\alpha},$$
$$g_2^a(\alpha) = \pi\left(\frac{R_m}{\sigma_0} - 1\right) + 2\sqrt{\alpha} + 2(1 + \alpha)\operatorname{arcctg}\sqrt{\alpha}, \tag{11}$$

where $\beta = w_0 / (HD\sigma_0)$ is a constant for the composite material.

The lower class boundary of equilibrium crack diameters follows from equality (9)

$$d \ge d_-, \quad \frac{2p}{\sigma_0}\left(\frac{d_-}{D}\right)^{1/2} = 2\sqrt{\beta}. \tag{12}$$

The upper class boundary follows from (9), and (10)

$$d \le d_+, \quad \frac{2p}{\sigma_0}\left(\frac{d_+}{D}\right)^{1/2} = g_1^a(\alpha_c), \tag{13}$$

where α_c is the limit (critical) value of reduced width of bridged zone, α.

The value α_c is the root of the equation

$$g_1^a(\alpha) = g_2^a(\alpha) \tag{14}$$

in the gap $D_f / D \le \alpha_c \le \alpha_0$, where D_f is the fiber diameter. The value $\alpha_0 = \beta / 4$ is a minimum of function $g_1(\alpha)$, and corresponds to value $d = d_-$. At $\alpha < D_f / D$ the bridging part is absent. Thus, if $\alpha_c < D_f / D$, the crack starts move only after breakage of all fibers. On other hand, in the case $\alpha_c > \alpha_0$ the lower bound of equilibrium crack values can not be described by this model.

Finally, the range of the equilibrium crack diameter can be expressed as

$$\kappa\lambda^a\left(\frac{R_f}{p}\right)^2 \le \frac{d}{D} \le \lambda^a\left(\frac{R_f}{p}\right)^2,$$
$$\sqrt{\lambda^a} = \frac{p\sqrt{d_+}}{R_f\sqrt{D}} = \frac{\sigma_0}{2R_f} g_1^a(\alpha_c), \quad \kappa = \frac{d_-}{d_+} = \frac{4\beta}{\left[g_1^a(\alpha_c)\right]^2}. \tag{15}$$

The crack is opened by two opposed point forces, P, applied at the center of the crack surfaces. Like in the previous case, we have the following equalities that correspond to equalities (9), and (10) based on the same assumptions $\Delta / a << 1$, $D / a << 1$:

$$\frac{4P}{\pi\sigma_0 d^2}\left(\frac{d}{D}\right)^{1/2} = g_1^a(\alpha), \tag{16}$$

$$\frac{4P}{\pi\sigma_0 d^2}\left(\frac{d}{D}\right)^{1/2} \le g_2^a(\alpha). \tag{17}$$

Hence,

$$\frac{4P}{\pi\sigma_0 d_+^2}\left(\frac{d_+}{D}\right)^{1/2} = 2\sqrt{\beta}, \quad \frac{4P}{\pi\sigma_0 d_-^2}\left(\frac{d_-}{D}\right)^{1/2} = g_1^a(\alpha_c). \tag{18}$$

As a result, we have two-sided inequality that is similar to (15):

$$\Lambda\left(\frac{P}{P_f}\right)^{2/3} \le \frac{d}{D} \le K\Lambda\left(\frac{P}{P_f}\right)^{2/3},$$
$$P_f = \frac{\pi D_f^2 R_f}{4}, \quad \Lambda = \left(\frac{D_f^2}{2D^2\sqrt{\lambda^a}}\right)^{2/3}, \quad K = \kappa^{-1/3}. \tag{19}$$

where P_f is a limit tensile force under which the fiber break.

Note, that formulas (12) – (15), (18), and (19) were obtained for fixed values of load parameters p and P. The alternative calculation at fixed values of crack diameter gives a range of load values, when the crack is equilibrium. Lower class boundary p_- (P_-) or upper class boundary p_+ (P_+) of these ranges are obtained from appropriate formulas (12), (13), or from (18), at substitution d for d_- or d_+, and p_- (P_-) or p_+ (P_+) for p (P), respectively. In this connection $\kappa = (p_- / p_+)^2$, and $\kappa = (P_- / P_+)^2$.

4. PENNY–SHAPED CRACK VS. CRACK UNDER PLANE DEFORMATION

As follows from the above, the critical size of the bridged part of a crack, Δ_c, is the same under remote loading, p, under effect of two concentrated forces, P, and under their combination. Hence, for the problem under consideration, quantity Δ_c, that defines the beginning of an unstable crack propagation, does not depend on the type of loading.

The same conclusion can be derived from Morozov et al. (1997) for the bridged crack under plane deformation. In addition, the quantity Δ_c is the same for plane deformation and for axisymmetric problems. Actually, this work results in an equation for α_c, which is equivalent to equation (14), i.e.,

$$g_1^p(\alpha) = g_2^p(\alpha) \tag{20}$$

where

$$g_j^p(\alpha) = \frac{2}{\pi} g_j^a(\alpha), \quad j = 1,2 .$$

Here and below superscript a corresponds to the axisymmetric problem, and superscript p corresponds to the plane problem.

Relations (9) and (10), (13) – (19) remain valid for the case of plane deformation as well. To apply these formulas to 2D-problem for a crack of length $L = 2l$ it is enough to replace g_j^a by g_j^p, d by L, and, moreover, in (12), (15) and (18) to replace β by $4\beta / \pi^2$. Like α_c the value κ is the same for both problems, i.e. $\kappa = d_- / d_+ = L_- / L_+$. At the same time we have following equality for the symbols λ^a, λ^p designated in (15) and (20):

$$\lambda^a / \lambda^p = \pi^2 / 4$$

Note, that for a crack under plane deformation point forces are uniformly distributed along a line, parallel to the crack front. In the above mentioned formulas symbol P for the plane case should be understood as the total force applied at the section of length L of median lines of the crack surfaces.

Let us to find a relationship between fracture toughness of the composite (i.e. the limit value of the stress intensity factor (SIF) for an unbridged crack) and the critical size of bridged part of the bridged crack for both problems. Asymptotics of the stress along the axis ρ, at $\rho \to a$, for the axisymmetric problem, and along axis x_1, at $x_1 \to l$, for the plane problem, can be written respectively as

$$\sigma_{33} = \frac{K^a}{\sqrt{2\pi(\rho - a)}} + O(1), \quad \sigma_{22} = \frac{K^p}{\sqrt{2\pi(x_1 - l)}} + O(1). \tag{21a}$$

Here $O(1)$ is the value of order of unity, factors K^a, K^p can be found from the solutions of the problems by means of the following formulas:

$$K^a = \lim_{\rho \to a} \sigma_{33}(\rho,\theta)\sqrt{2\pi(\rho - a)}, \quad K^p = \lim_{x_1 \to l} \sigma_{22}(x_1, 0)\sqrt{2\pi(x_1 - l)}. \tag{21b}$$

After substitution of the right-hand part of formula (6) into the first equality in (21b), and the solution of the plane problem from Morozov et al. (1997) (taking into account the effect of the point forces at center of the crack) into second equality. We obtain

$$K^a = K_1^a - 2\sigma_0\sqrt{\frac{a^2 - b^2}{\pi a}}, \quad K^p = K_1^p - \frac{2\sigma_0\sqrt{l}}{\sqrt{\pi}} \arccos(1 - \Delta / l). \tag{22}$$

Here

$$K_1^a = \frac{2p\sqrt{a}}{\sqrt{\pi}} + \frac{P}{\pi a\sqrt{\pi a}}, \quad K_1^p = p\sqrt{\pi l} + \frac{P}{2l\sqrt{\pi l}}$$

are SIFs for unbridged cracks.

To find the limit values of the last factors, i.e. K_{1c}^a and K_{1c}^p, substitute first relation from (21) into criterion (3), and second relation from (21) into analogous condition (17) of Morozov et al. (1997). Then, taking into account (22) and the asumptions $\Delta << a$, $\Delta << l$, we obtain the following equalities:

$$K_{1c}^a = K_{1c}^p \equiv K_{1c} = \left(\pi R_m + 4\sigma_0\sqrt{\alpha_c}\right)\sqrt{\frac{D}{2\pi}}. \tag{23}$$

Thus, fracture toughness K_{1c} of an aligned composite is determined by three material parameters: tensile strength of the matrix, R_m, fracture zone size, D, and critical size of bridged part of the crack, $\Delta_c = \alpha_c D$. If $\Delta_c = 0$ the fibers do not resist explicitly to the crack growth, but oppose crack opening as, in general, $\Delta_0 = \alpha_0 D \neq 0$. Nevertheless, in this case K_{1c} will still differ from the matrix fracture toughness if the reinforcement affects the fracture zone size, i.e. if $D \neq D_m$. The reinforcement influences also the limit crack size, because, according to formulas (13) and (18), this size depends on value D.

5. EXAMPLE

Let us consider force-separation law $\sigma(w)$ from Budiansky et al. (1995), where the problem of elastic-plastic drawing of the fiber from the matrix was solved:

$$\frac{\sigma}{\sigma_A} = \begin{cases} \dfrac{w}{D_f}, & w \le w_B, \\ \sqrt{\dfrac{w - w_B}{D_f} + \left(\dfrac{w_B}{D_f}\right)^2}, & w \ge w_B, \end{cases} \qquad (26)$$

Here

$$\sigma_A = \frac{R_f}{D_f\sqrt{(w_0 - w_B)D_f + w_B^2}}.$$

The instances of materials with such a pattern of fiber pull out are silicon carbide fibers in a silicon carbide matrix (SiC/SiC) and the same fibers in a calcium-aluminosilicate glass ceramic matrix (SiC/CAS). Their data, based on Budiansky et al. (1995), are shown in Table 1. One observes slightly anisotropic material properties: Young's modulus ratio $E_2/E_1 \approx 0.95$ for both materials.

The characteristics of cracking based on present model are shown in Table 2. The measures Δ_0 and Δ_c are the width of crack bridged part as the crack diameter is equal lower and upper class boundary of equilibrium cracks range (these values are divided by fiber diameter D_f). Parameters λ^a, κ, according to (15) and (19), describe the critical equilibrium state of the crack. K_{1c} is the fracture toughness (in $MPa\sqrt{m}$), given by (23). Values of all parameters are given for different sizes of matrix fracture zone (4).

By varying value w_0 but conserving all other parameters of the composite, we have the relations $\alpha_c, \lambda^a, \kappa$ vs. w_0 from equations (14), and (15). These parameters for material SiC/SiC at $D = D_m$ are shown in Figs. 4 and 5. The values of param eters which correspond to the real value of w_0 (see Table 1) are marked by arrows. Partly λ^a, κ are given in Table 2. Solid lines correspond to penny-shaped cracks, dashed lines

Table 1. Data of two composite materials.

Material	E_f	E_m	ν_f	ν_m	R_f	σ_0	R_m	c	w_0	D_f	D_m
	GPa				GPa				μm		
SiC/SiC	200	300	0.4	0.4	2	0.53	2.83	0.4	1.03	7	0.25
SiC/CAS	200	100	0.4	0.4	2	0.53	0.96	0.4	2.04	7	1.79

Table 2. Parameters of equilibrium and critical states of a bridged crack vs. the size of fracture zone D.

Material	D	Δ_0 / D_f	Δ_c / D_f	λ^a	κ	K_{1c}
SiC/SiC	D_m	67,14	57,86	541.7	0.994	18.83
	$2D_m$	67,14	54,21	272.4	0.988	19.01
	$4D_m$	67,14	49.51	137.9	0.976	19.31
SiC/CAS	D_m	71,6	66,03	79.50	0.998	19.93
	$2D_m$	71,6	63,83	39.81	0.997	20.29
	$4D_m$	71,6	60,77	19.97	0.994	20.80

correspond to the crack under plane deformation. Analogous curves for material SiC/CAS agree qualitatively with lines represented in Figs. 4-5.

According to formulas (13), (18) the relations of ultimate load p_+, P_+ vs. penny-shaped crack diameter d (solid line) or 2D–crack length L (dashed line) for material SiC/SiC at $D = D_m$ are plotted in Figs. 6. Analogous lines for p_-, P_- are practically identical since $\kappa \approx 1$ (see Table 2).

This identity means, that for a crack in a critical state not only the critical size, Δ_c, of its bridged part in the, but any value range $\Delta_c \le \Delta \le \Delta_0$ is possible. Therefore, at given load the crack of critical size will not start until the size of its bridged part, Δ, becomes equal to critical value Δ_c.

Note, that although these materials differ appreciably in some parameters (see Table 1), the ultimate loads for the cracks in composite SiC/CAS are only 2.5% higher, then ones for SiC/SiC. Consequently, lines in Figs. 6 can be referred to composite SiC/CAS as well (taking into account mentioned difference in ultimate loads).

6. DISCUSSION

According to Table 2, at limit state the width of the bridged part of a circular crack is equal approximately $58D_f$ for SiC/SiC and $66D_f$ for material SiC/CAS, when $D = D_m$. As in our model $\Delta / a < <1$ the crack size at the limit state should satisfy the condition $d_+ >> 116D_f$ ($d_+ >> 812$ μm) for SiC/SiC and $d_+ >> 132D_f$ ($d_+ >> 924$ μm) for SiC/CAS. Hence, for given materials our model a llows us to evaluate the limit size of the crack if it is more than one millimeter. Therefore diagrams in Figs. 6 are truncated to the left.

This restriction can be moved to the left by reducing of value β. Indeed, if β decreases the root of equation (14) decreases as well, and by virtue of a monotony of function $g_2(\alpha)$ the value $g_2(\alpha_c)$, defining the value d_+, decreases. The value β is directly proportional to parameter w_0 and inversely proportional to parameters H, D, and σ_0. These relations point out the way of reduction of β. In particular, Fig. 4a shows how the root α_c of the parameter w_0 changes, when affected by the fiber stiffness and the adhesive strength.

The problem of determination of another important characteristic of fracture, i.e. the size of fracture zone, D, should be discussed especially. Speaking strongly, equality $D = D_m$ is justified for crack in isotropic body, a material is the same as the matrix of composite. For fiber–reinforced material value D can be different, because the accumulation of the defects in the matrix occurs in the neighborhood of the fibers, which influence the fracture process.

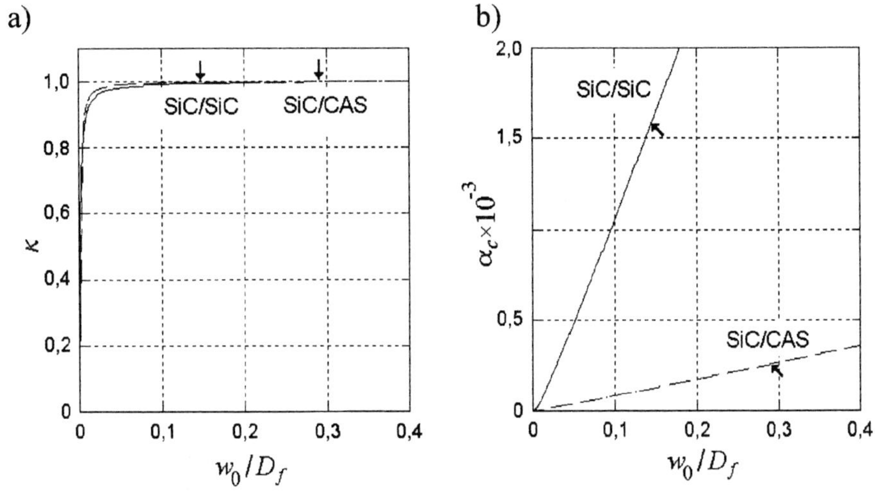

Fig. 4. The dimensionless critical width, α_c, of a bridged crack part (a) and the ratio, κ, of the minimal crack diameter to the critical one (b) vs. critical crack opening displacement, w_0, under which fibers are broken.

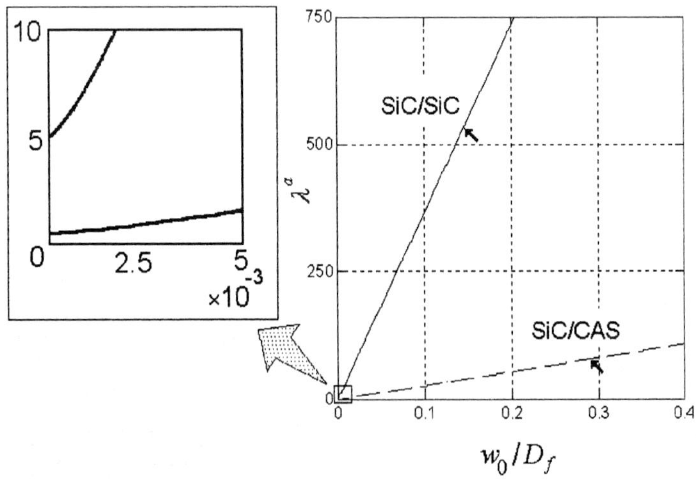

Fig. 5. Dimensionless fracture toughness squared, λ^a, vs. critical crack opening displacement, w_0, for two ceramics.

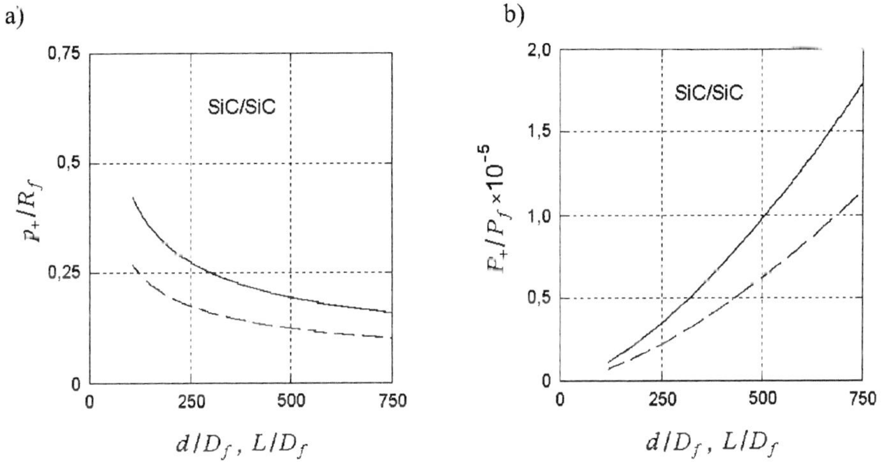

Fig. 6. Ultimate remote load, p_+, and ultimate point forces, P_+, vs. diameter, d, of a circular crack (solid lines) or length, L, of a crack under plane deformation (dashed lines).

Thus, D is simultaneously a characteristic revealing of matrix damage and a characteristic of the heterogeneity of the composite. The simplest way to take into account of an influence of material heterogeneity on matrix damage is the usage of the relation in form (4).

Using the equality $D = D_m$ is stronger motivated for SiC/SiC. Actually, since the matrix fracture takes place between the rows of the fibers, the determining factor for the estimation of the structure influence is ratio D_m / u, where u is the distance between neighbor rows of the fibers. The smaller this ratio the smaller the influence of the fibers on the matrix fracture is. At hexagonal arrangement of the fibers the distance between fiber rows is defined from equality $u^2 = \pi D_f^2 \sqrt{3} / (8c)$. It follows that (see Table 1) $u = 9{,}13$ μm for both composites, $D_m / u = 0{,}027$ for SiC/SiC and $D_m / u = 0{,}2$ for SiC/CAS. Thus, for material SiC/SiC we have $D_m << u$.

Note however, that although ratio D_m / u for material SiC/CAS is much greater than that for SiC/SiC, the dependence of the parameters in Table 2 on value D is the same for both materials.

From Table 2 it follows that:

1. Even a considerable change of the size of the fracture zone, D, does not affect the product λD appreciably, i.e. the critical size of the crack (d_+ or L_+) at fixed load. The fracture toughness, K_{1c}, responds similarly to changes of D.
2. For any value D the equality $d_+ / L_+ = \pi^2 / 4 \approx 2{,}47$ may be used.
3. The influence of value D on the critical size of the bridged crack part, Δ_c, is the most considerable. As the value D increases, the value Δ_c decreases, and therefor, difference $\Delta_0 - \Delta_c$ increases. Simultaneously $\kappa = d_- / d_+$ practically does not change.

Acknowledgments

This research was supported by Russian Foundation for Fundamental Research (99-01-00673 and 00-01-00484).

References

1. Grekov, M.A. and Morozov, N.F. (2000) Disk-shaped equilibrium cracks, *Prikl. Mat. Mekh.*, **64**, No. 1, pp. 172-175 (in Russian). *J. Appl. Math. Mech.*, **64**, No. 1, pp. 169-172 (English translation).
2. Novozhilov, V.V. (1969a) On the necessary and sufficient criterion for brittle strength, *Prikl. Mat. Mekh.*, **33**, No. 2, pp. 212-222 (in Russian). *J. Appl. Math. Mech.*, **33**, No. 2, pp. 201-210 (English translation).
3. Novozhilov, V.V. (1969b) On the foundation of a theory of equilibrium cracks in elastic solids, *Prikl. Mat. Mekh.*, **33**, No.5, pp. 797-812 (in Russian). *J. Appl. Math. Mech.*, **33**, No. 5, pp. 777-790 (English translation).
4. Thomson, R.and Hsieh and C., Rana, V. (1971) Lattice trapping of fracture cracks, *J. Appl. Phys.*, **42**, No 8, pp. 3154–3160.
5. Morozov, N.F. and Paukshto M.V. (1988) On 'lattice' trapping, *Dokl. AN SSSR.*, **299**, No. 2, pp. 323-325 (in Russian). Sov. Phys.–Dokl., **33**, No. 3, pp. 235-236. (English translation).
6. Morozov, N. and Paukshto, M. (1991) On the crack simulation and solutions in the lattice, *Trans. ASME J. Appl. Mech.,* **58**. No 3, pp. 290–292.
7. Morozov, N. and Paukshto, M. and Ponikarov, N. (1997) On the problem of equilibrium length of bridged crack, *Trans. ASME. J. Appl. Mech.,* **64**, No. 2, pp. 427–430.
8. Nemat-Nasser, S. and Hori, M. (1987) Toughening by partial or full bridging of crack in ceramics and fiber reinforced composites, *Mechanics of Materials*, **6**, pp. 245–269.
9. Movchan, N.V. and Willis, J.R. (1998) Penny-shaped crack bridged by fibers, *Quarterly of Applied Mathematics.*, **56**, No. 2, pp. 327–340.
10. Budiansky, B. and Evans, A.G. and Hutchinson, J.W. (1995) Fiber-matrix debonding effects on cracking in aligned fiber ceramic composites, *Int. Journal of Solids and Structures*, **32**, No. 3/4, pp. 315–328.
11. Morozov, N.F. (1984) *Mathematical Problems of Cracks Theory,* Nauka, Moskva (in Russian).
12. Panasyuk, V.V. (1968) *The Limit Equilibrium of Brittle Bodies with Cracks*, Naukova Dumka, Kiev (in Russian).
13. Fabrikant, V.I. (1990) Complete solution to some mixed boundary value problems in elasticity, *Advances in Applied Mechanics.* Eds. J.Hutchinson and T.Wu. Boston. Acad. Press, **27**, pp. 153–223.
14. Christensen, R.M. (1979) *Mechanics of composite materials*, John Wiley & Sons Inc.

IDENTIFICATION OF STRESS FROM STRAIN FOR BODY WITH NOT FULLY IDENTIFIED CONSTITUTIVE RELATIONS

Muneo Hori
Earthquake Research Institute, University of Tokyo, Yayoi, Tokyo 113-0032, Japan

Toshihiro Kameda
Department of Structural Engineering, Tsukuba University, Tenodai, Tsukuba, 305-8577, Japan

Abstract This paper presents a new inversion method to identify a distribution of stress from a distribution of measured strain for a body whose constitutive relations are not fully known. The present method yields well-posed boundary value problems for unknown stress. Local constitutive relations are then determined from relations between the measured strain and the predicted stress. This procedure is different from an ordinary analysis of finding constitutive relations, which is usually formulated as the parameter identification by assuming a form of the constitutive relations. The rigorous formulation of the inversion method is briefly presented. Also, three examples to which the present method has been applied are shown. The validity and the usefulness of the method are discussed in view of the results obtained.

1. INTRODUCTION

The identification of constitutive relations is of primary importance in analyzing heterogeneous, anisotropic or non-linear materials; see, for instance, Bui (1994); see also Hori *et al.* (1999) for a concise list of related references. Assuming a form of the constitutive relations and applying suitable parameterization to the form, we can formulate an inverse problem to find the constitutive relations. Measured displacement and boundary traction are used as input data. While it is straightforward, this inversion method has the following two drawbacks:

P. Ståhle and K.G. Sundin (eds.), IUTAM Symposium on Field Analyses for Determination of Material Parameters - Experimental and Numerical Aspects, 217–229.

1 there are cases when the validity of the assumed form is not fully verified.

2 finding suitable parameters results in a non-linear optimization problem which cannot be solved easily.

The first drawback is serious if standard material tests cannot be applied to a target body consisting of an unknown material due to, say, its extremely small or large size.

Recently, the authors are proposing a new inversion method to identify the local constitutive relations; see Hori *et al.* (1999) and Hori and Kameda (2000). The key feature of the proposed method is that it is aimed at identifying a local stress distribution, not local constitutive relations. Thus, the form of the constitutive relations does not have to be assumed and no parameterization is needed. While the method is applicable only to a body in a two-dimensional state, the method can yield well-posed boundary value problems for an unknown stress distribution. Solving these problems, we can obtain stress from strain even though the constitutive relations, heterogeneous, anisotropic or non-linear, are not fully known. The data required for the analysis are a distribution of measured strain and boundary traction. Also, information about one (not full) constitutive relation is needed. When the stress distribution is predicted, it is straightforward to find local constitutive relations by comparing the local stress with the corresponding strain.

In this paper, we present the proposed inverse method, which will be called stress inversion here. Some emphases will be put on examples to which the method has been applied. After the basic formulation is briefly shown in Section 2, three examples are presented in Section 3, namely, the numerical simulation of an elasto-plastic material sample, the model experiment of a gelatin sample, and the analysis of the Global Positioning System (GPS) data of the deformation of the Japanese Islands. These examples demonstrate the validity and the usefulness of the stress inversion.

To minimize mathematical expression, we use index notation and employ summation convention. Also, we denote partial differentiation by a comma followed by a suffix.

2. FORMULATION OF STRESS INVERSION METHOD

We consider a thin body, V, which is in plane stress state. The x_1 and x_2 coordinates are taken on the surface S. Displacement u_i is measured on S and traction t_i is measured along the boundary ∂S; see Fig. 1.

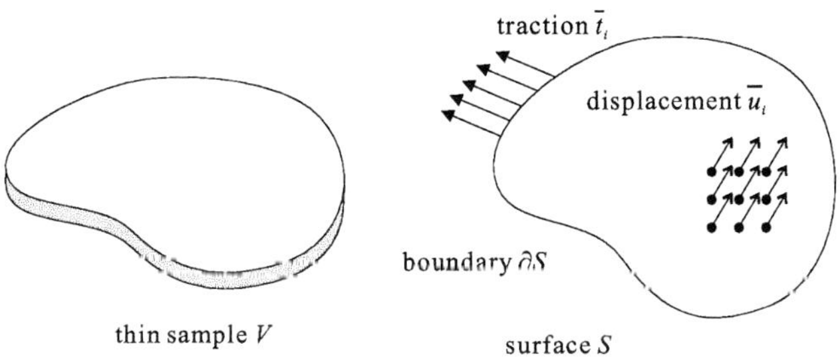

Figure 1 Body in plane stress state.

For simplicity, we assume that a body is elasto-plastic and that the two-dimensional plastic strain is incompressible. That is, strain rate $\dot{\epsilon}_{ij}$ is decomposed into elastic and plastic parts, $\dot{\epsilon}_{ij} = \dot{\epsilon}^{\mathrm{e}}_{ij} + \dot{\epsilon}^{\mathrm{p}}_{ij}$, and $\dot{\epsilon}^{\mathrm{p}}_{ij}$ satisfies

$$\dot{\epsilon}^{\mathrm{p}}_{ii} = 0. \tag{1.1}$$

Now, we consider three components of stress rate, $\dot{\sigma}_{ij}$, which always satisfy two equation of equilibrium, i.e.,

$$\dot{\sigma}_{ij,i} = 0 \tag{1.2}$$

The three equations given by Eqs. (1.2) and (1.1) form coupled differential equations for unknown stress rate. To simplify expression, however, we introduce *eigenstress* rate as $\dot{\sigma}^{*}_{ij} = -c^{\mathrm{o}}_{ijkl}\dot{\epsilon}^{\mathrm{p}}_{kl}$ where c^{o}_{ijkl} is the elasticity of the material. The governing equations for $\dot{\sigma}^{*}_{11}(= -\dot{\sigma}_{22})$ and $\dot{\sigma}_{12}$ are now derived from Eq. (1.2), as follows:

$$\begin{aligned} \dot{\sigma}^{*}_{11,1} + \dot{\sigma}^{*}_{12,2} &= -\dot{\sigma}^{\mathrm{a}}_{i1,i}, \\ \dot{\sigma}^{*}_{12,1} - \dot{\sigma}^{*}_{11,2} &= -\dot{\sigma}^{\mathrm{a}}_{i2,i}, \end{aligned} \tag{1.3}$$

where $\dot{\sigma}^{\mathrm{a}}_{ij}$ is apparent stress computed from measured strain rate as $\dot{\sigma}^{\mathrm{a}}_{ij} = c^{\mathrm{o}}_{ijkl}\dot{\epsilon}_{kl} = c^{\mathrm{o}}_{ijkl}\dot{u}_{k,l}$. Measured traction rate prescribes boundary conditions, as

$$\begin{aligned} n_1\dot{\sigma}^{*}_{11} + n_2\dot{\sigma}^{*}_{12} &= \dot{t}_1 - n_i\dot{\sigma}^{\mathrm{a}}_{i1}, \\ n_1\dot{\sigma}^{*}_{12} - n_2\dot{\sigma}^{*}_{11} &= \dot{t}_2 - n_i\dot{\sigma}^{\mathrm{a}}_{i2}, \end{aligned} \tag{1.4}$$

where n_i is the unit normal of ∂S.

Rigorously speaking, we do not need to have two boundary conditions in Eq. (1), in order to solve Eq. (1) for $\dot{\sigma}^{*}_{11}$ and $\dot{\sigma}^{*}_{12}$. If two traction components are given, the coupled boundary value problems are easily decomposed; for instance, $\dot{\sigma}^*_{11}$ satisfies

$$\begin{cases} \dot{\sigma}^*_{11,ii} = -\dot{\sigma}^{\mathrm{a}}_{11,11} + \dot{\sigma}^{\mathrm{a}}_{22,22} & \text{on } S, \\ \dot{\sigma}^*_{11} = n_1\dot{t}_1 - n_2\dot{t}_2 - ((n_1)^2\dot{\sigma}^{\mathrm{a}}_{11} - (n_2)^2\dot{\sigma}^{\mathrm{a}}_{22}) & \text{along } \partial S. \end{cases} \tag{1.5}$$

As is seen, the governing equation is the Poisson equation and the boundary condition is the Direchlet type. Therefore, the solution of Eq. (1.5) is unique and stable. A similar boundary value problem is derived for $\dot{\sigma}^*_{12}$. Using the solution of the decoupled boundary value problems, $\dot{\sigma}^*_{ij}$, we can obtain the stress rate, as follows:

$$\dot{\sigma}_{ij} = \dot{\sigma}^{\mathrm{a}}_{ij} + \dot{\sigma}^*_{ij}. \tag{1.6}$$

It should be emphasized that the stress inversion method is applicable to any plastic constitutive relations, provided that the assumed condition, Eq. (1.1), holds. This means that the stress rate for a non-linear body can be obtained by solving the linear boundary value problems. The constitutive relations are easily found by plotting the stress rate with respect to the strain rate, once the stress rate is predicted.

The key point of the stress inversion is the fact that the three stress rate components satisfy the two equations of equilibrium. Only one condition is needed to determine the three components, and Eq. (1.1) is an example of the condition. The condition that stress is in equilibrium is always satisfied, if Airy's stress function is used. That is, we can express equilibrating stress components in terms of Airy's stress funtion, denoted by a, as

$$\{\sigma_{11}, \sigma_{22}, \sigma_{12}\} = \{a_{,22}, a_{,11}, -a_{,12}\}. \tag{1.7}$$

Usually, a is determined such that the compatibility condition of strain is satisfied when constitutive relations are known; for instance, a satisfies a bi-harmonic equation for an isotropic elastic material. If one relation between strain and stress is known, a can be determined by measuring a distribution of strain. For instance, if the two-dimensional stress rate $\dot{\sigma}_{11} + \dot{\sigma}_{22}$ is given by a function of strain rate as $f(\dot{\epsilon}_{ij})$, we have

$$\dot{a}_{,ii} = f(\dot{\epsilon}_{kl}). \tag{1.8}$$

In this form, the condition assumed by Eq. (1.1) is rewritten as $f = c^{\mathrm{o}}_{iikl}\dot{\epsilon}_{kl}$. We can solve Eq. (1.8) for $\dot{a}$, using the boundary conditions $\dot{t}_1 = n_1\dot{a}_{,22} - n_2\dot{a}_{,12}$ and $\dot{t}_2 = -n_1\dot{a}_{,12} + n_2\dot{a}_{,11}$, which can be rewritten as

$$n_i\dot{a}_{,i} = -n_1\dot{r}_1 + n_2\dot{r}_2 \tag{1.9}$$

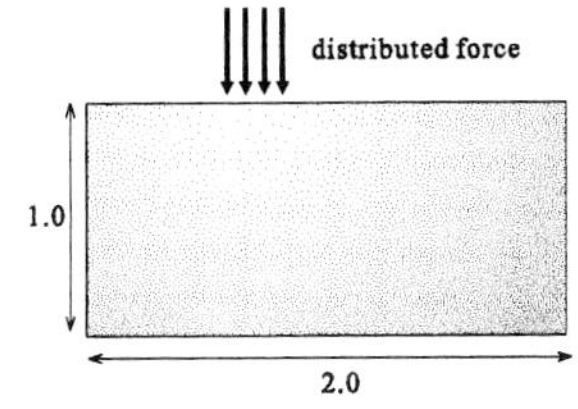

Figure 2 Numerical simulation of material sample.

with $\dot{r}_i$ being the resultant rate force that is obtained by integrating $\dot{t}_i$ along the boundary, $\dot{r}_i(\mathbf{x}) = \int^{\mathbf{x}} \dot{t}_i \mathrm{d}\ell$; see Hori and Kameda (2000). The governing equation (1.8) is the Poisson equation and the boundary conditions (1.9) are the Neumann type. Hence, the solution of Eq. (1.8) and (1.9) is unique and stable.

It should be noted that stress rate given by solving the boundary value problem of $\dot{a}$ is identical with that given by solving the decoupled boundary value problem of $\{\dot{\sigma}^*_{11}, \dot{\sigma}^*_{12}\}$. The following two are the major differences: 1) solving $\{\dot{\sigma}^*_{11}, \dot{\sigma}^*_{12}\}$ need to take higher derivative of measured displacement; and 2) the second derivative of $\dot{a}$ is needed to obtain $\dot{\sigma}_{ij}$. Depending on the accuracy of measuring displacement and the numerical analysis method to solve the boundary value problem, we can choose either $\dot{a}$ or $\{\dot{\sigma}^*_{11}, \dot{\sigma}^*_{12}\}$.

3. APPLICATION OF STRESS INVERSION METHOD

This section briefly presents the results of applying the stress inversion to example problems, in order to verify the validity and to demonstrate the usefulness; Hori *et al.* (1999), Hori and Kameda (2000) and Hori *et al.* (2000) provide more detailed explanations of the examples problems.

3.1. Numerical simulation

The target of the numerical simulation is a rectangular sample consisting of an unknown elasto-plastic material; see FIg. 2. It is assumed that the elasticity is isotropic with Young's modulus E and Poisson's ratio ν. The plastic property is unknown, although information about Eq. (1.1) is provided. In the simulation, we use a von-Mises yield function, $f = \tau - \tau_y$, with τ and τ_y being the maximum shear stress and the yield stress, and generate input data of the stress inversion; the values of the material constants are set as $E = 1.0$[MPa], $\nu = 0.3$ and $\tau_y = 0.001$[MPa], respectively. In the numerical simulation, dis-

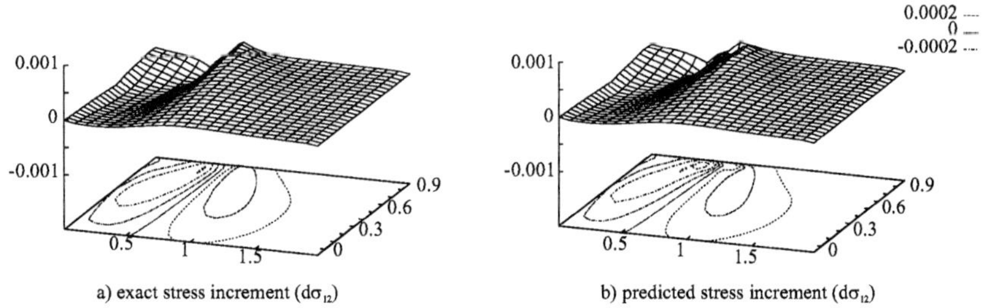

Figure 3 Comparison of stress distribution.

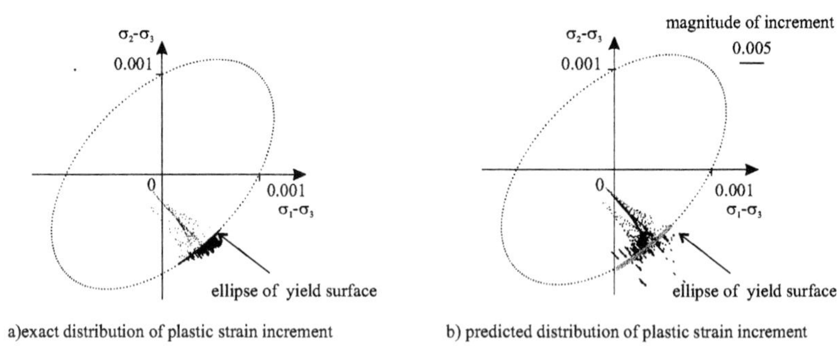

Figure 4 Comparison of yield locus.

placement and traction are first computed by using the finite element method. The stress inversion then solves the decoupled boundary value problems for the eigen-stress and obtains the stress increment and the stress. Other conditions of the simulation such as boundary conditions are summarized in Hori *et al.* (1999).

As a typical example of the stress inversion, Fig. 3 shows the distribution of exact and predicted $\dot{\sigma}_{12}$. As is seen, the agreement is satisfactory. Indeed, the stress concentration at the loading is well reproduced by the predicted $\dot{\sigma}_{12}$. We then compute plastic strain rate, subtracting the elastic strain rate from the measured (total) strain rate, in order to find the plastic constitutive relations. Figure 4 plots the plastic strain in the two-dimensional plane of the principle stress. The exact inelastic strain rate forms an edge of an ellipse that corresponds to the yield surface of von Mises. While the predicted inelastic strain rate forms similar edge less clearly, we can read the yield stress as $\tau_y = 0.001 \pm 0.0002$[MPa], which is close to the exact value of 0.001[MPa]. These two figures obtained by the numerical simulation, Fig. 3 and 4, are good examples that verify the validity of the stress inversion.

3.2. Model experiment

The target of the model experiment is a gelatin sample; see Fig. 5. The sample is disk-shaped, with radius 30[cm] and thickness 1[cm], and subjected to the torsional shearing at the bottom by the rotating a coaxial circular part of radius 15[cm], with the peripheral site being fixed. Displacement is measured by using image analysis; markers forming grid-pattern are put on the surface, the deformation processes are recorded in a video tape, and the movement of all markers is computed by analyzing their location in each image of the record. Examples of the video image and the analyzed maximum shear strain distribution are shown in Figs. 6 and 7, respectively. The formation of shear cracks and the strain concentration are observed. Hori and Kameda (2000) provides detailed explanations of the experiment such as boundary conditions and the image analysis procedures.

As a typical example, Fig. 8 plots relations of σ_{11}-ϵ_{11} and σ_{12}-ϵ_{12}. The stress are obtained by assuming Eq. (1.1) and applying the stress inversion. These relations are for distinct two points. While Fig. 8 suggests the heterogeneity, some common constitutive relations, which show the homogeneity, are obtained for this material. In Fig. 9, the relations of the maximum shear stress and strain, τ and γ, are plotted; A~E are the five points shown in Fig. 6. For C and D where the crack passes, γ increases even though τ attains an almost constant value. For other

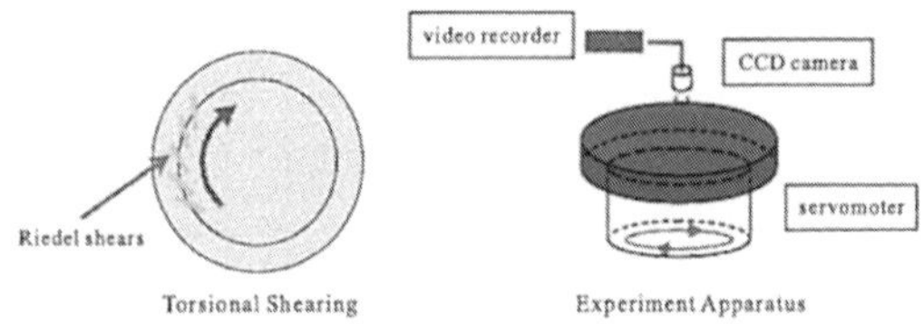

Figure 5 Schematic View of Model Experiment

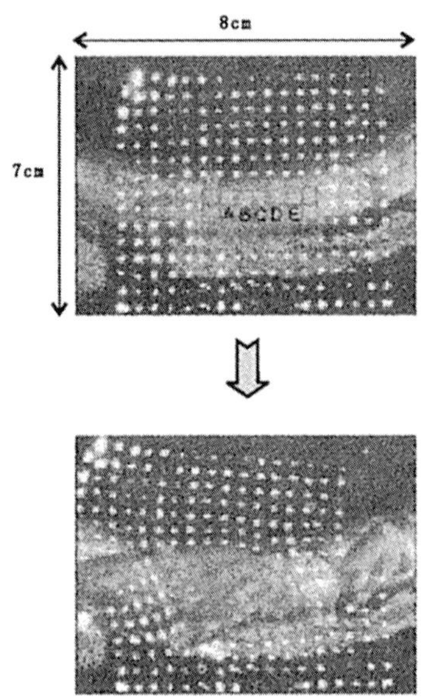

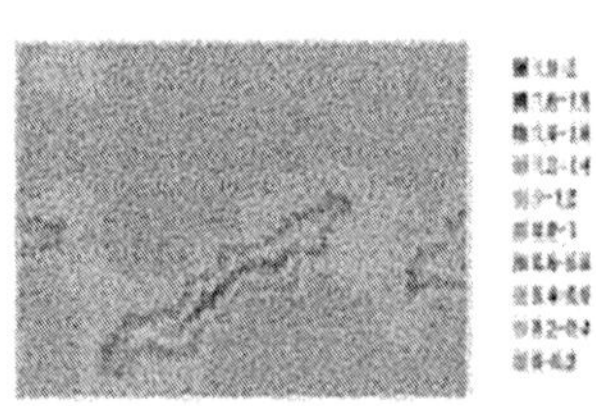

Figure 7 Maximum shear strain distribution obtained by image analysis.

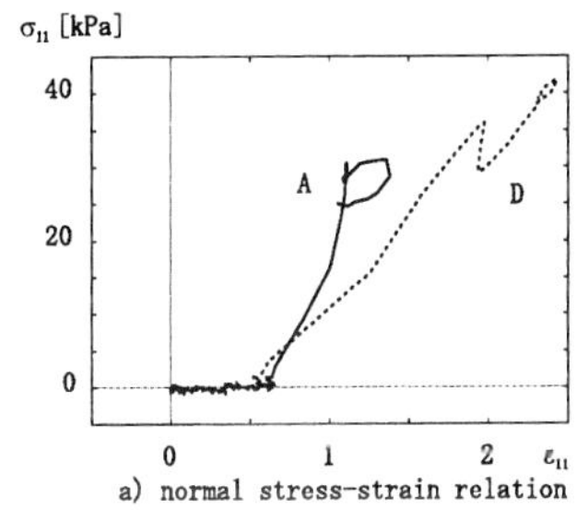

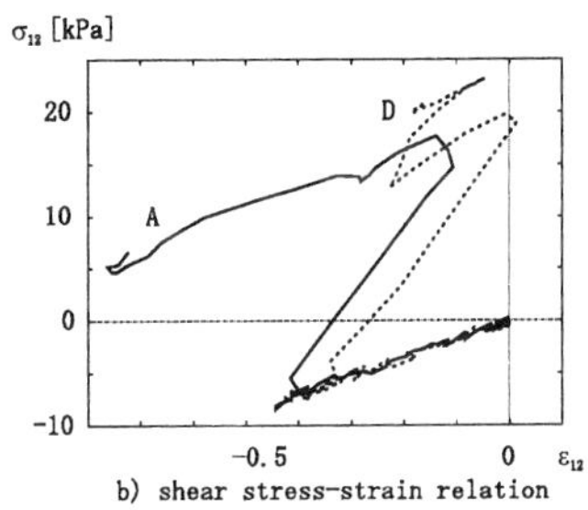

Figure 8 Strain-Stress relations of two points.

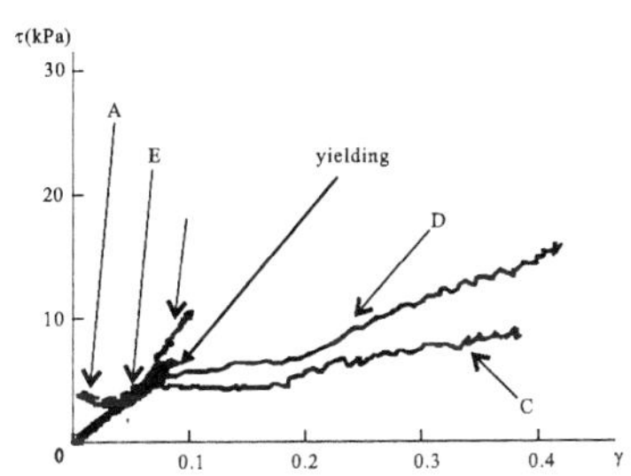

Figure 9 Relation between maximum shear stress and strain of five points.

points, τ increases almost linearly with respect to γ, probably because no failure takes place near these points. These relations are different from the overall shear strain and stress relations that are measured by the amount of the rotation and the applied torque. Provided that the assumed equation (1.1) holds, the present model experiment is an example of the usefulness of the stress inversion.

3.3. GPS data analysis

The GPS data analysis uses displacement increment of the Japanese Islands measured by the nationwide GPS array that consists of around 1000 stations. Although the crustal stress state varies vertically, the plane stress state can be assumed for the increment deformation during a short period of the GPS measurement. Thus, the stress inversion is applicable. The GPS data are filtered through the least square prediction to get rid of measurement noises; see El-Fiky (1999). It is assumed that the elastic property is isotropic, with Young's modulus and Poisson's ratio being $E = 1$[MPa] and $\nu = 0.25$, and that boundary tractions are zero; see also Hori *et al.* (2000) for the explanation of the analysis conditions.

For the measured strain increment and the predicted stress increment, we plot the distribution of the first and second invariants, $\{\dot{\epsilon}, \dot{\sigma}\}$ and $\{\dot{\gamma}, \dot{\tau}\}$, and the distribution of the direction of the maximum shear

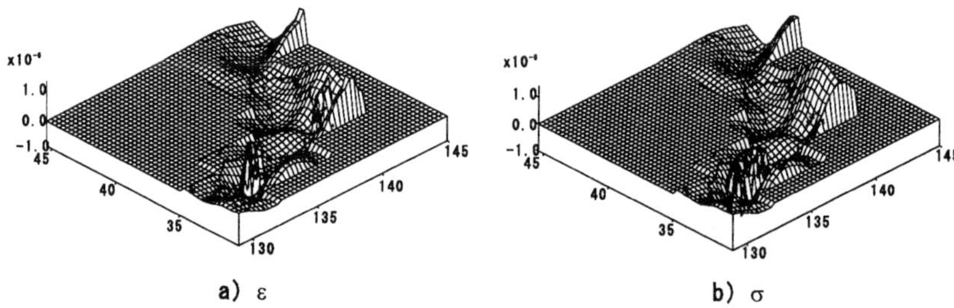

Figure 10 Distribution of $\dot{\epsilon}$ and $\dot{\sigma}$.

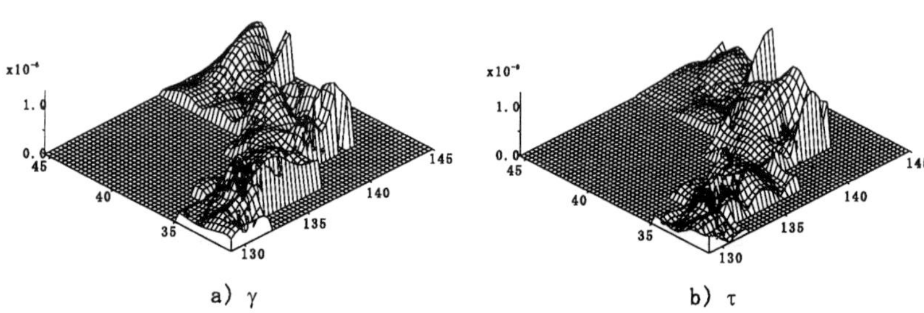

Figure 11 Distribution of $\dot{\gamma}$ and $\dot{\tau}$.

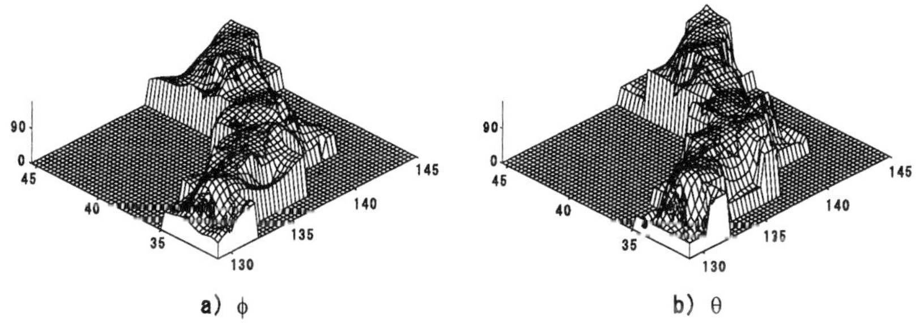

Figure 12 Distribution of $\phi - \theta$

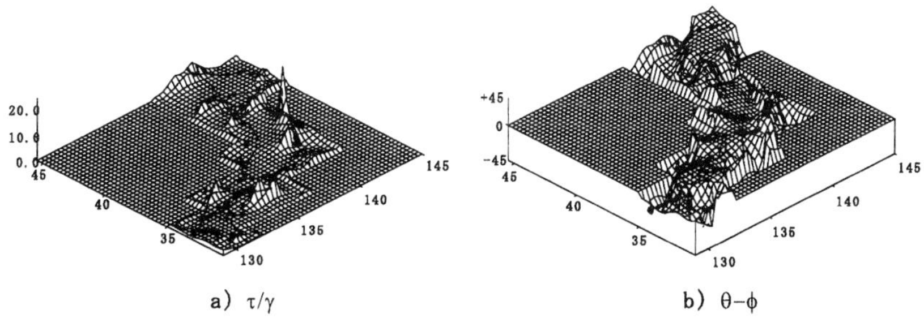

Figure 13 Regional heterogeneity and anisotropy.

component, $\{\theta, \phi\}$, in Figs. 10, 11 and 12, respectively. The distribution of the predicted stress increment measure differs from that of the corresponding measured strain increment measure; for instance, Figs. 10 and 11 show that some regions have a higher concentration of the stress increment compared with the strain increment; also a locally large difference in the maximum shear component direction is observed in Fig. 12. In order to find regional constitutive relations, we plot in Fig. 13 the shear stiffness given by $\dot{\tau}/\dot{\gamma}$ and the difference in the maximum shear direction given by $\phi - \theta$. These figures show the regional heterogeneity and anisotropy. While it is impossible directly to measure regional constitutive relations, they have to be compared with known geological data to examine the validity. However, these figures obtained present the potential usefulness of the stress inversion to analyze the GPS data.

4. CONCLUDING REMARKS

The stress inversion is presented in this paper. This inversion method provides boundary value problems which are prescribed by using measured data and a certain relation between stress and strain. The validity and the usefulness are shown in three example problems. While further verification is needed, we are expecting that the stress inversion is a versatile tool to identify local constitutive relations for material samples.

Acknowledgments

This research is supported partially by Grant-in-Aid for Scientific Research, the Ministry of Education, Science, Sprots and Cultuer and partially by Japan Scienc and Technology Corporation.

References

Bui, H.D. (1994). *Inverse Problem in the mechanics of materials: An introduction*, CRC Press, New York.

El-Fiky, G.S., and Kato, T. (1999). Interplate coupling in the Tohoku district, Japan, deduced from geodetic data inversion, *J. Geophys. Res.*, **104**, B9, 20361-20377.

Hori, M., Kameda, T. and N. Hosokawa, N. (1999). Formulation of identifying material property distribution based on equivalent inclusion method, *Structural Eng./Earthquake Eng.*, JSCE, **16**, 1, 21-30.

Hori, M. and Kameda, T. (2000). Inversion of stress from strain without fully knowing constitutive relations, *J. Mech. Phys. Solids* (in review process).

Hori, M., Kameda, T. and Kato, T. (2000). Application of Inversion Method to GPS Network for Estimating Stress Increment In Japan, *Geophys. J. Int.* (in review process).

Nemat-Nasser, S. and Hori, M. (1998). *Micromechanics: Overall Properties of Heterogeneous Materials*, Second Edition, North-Holland, London.

EVALUATION METHOD OF INTERFACE STRENGTH BETWEEN THIN FILMS IN ADVANCED LSI

Takayuki Kitamura, Tadahiro Shibutani and Takashi Ueno
Kyoto University, Graduate School of Engineering, Kyoto, 606-8501, Japan

Abstract. This paper aims to develop a method for evaluating the interface strength between thin films on a silicon substrate of LSI. Although several methods have been proposed, plastic deformation and fracture of thin metal film bring about ambiguity in the magnitude of the interface strength obtained. Especially, it is important to determine the criterion of delamination crack initiation from the edge where the stress concentrates due to the mismatch of elastic deformation. In this study, an evaluation method is proposed, and the validity is examined by experiments on Cu (conductor metal)/TaN (barrier metal) interface on a silicon substrate. The stress intensity at the delamination crack initiation is successfully analyzed by the boundary element method (BEM).

1. INTRODUCTION

In the electronics industry, the reduction of component size has been intensely pursued, and the thickness of LSI device is reaching less than 1mm. Since the sub-micron components consist of multi-layered films made of ceramics, intermetallic compounds and metals, the delamination at the interface becomes one of the major failure modes in the processing and in service of LSI. Therefore, it is important in terms of reliability to evaluate the interface strength between thin films. Especially, since the delamination crack usually starts at the edge of thin films due to the stress concentration from the mismatch of deformation, it is important to evaluate the free edge effect [1].

Many evaluation methods for the interface strength have been developed, and the popular methods are schematically shown in Fig.1 [2-5]. Although the results give a qualitatively good measure for the adhesive strength of film on the substrate, the results show large scatter and the mechanical meaning is not clear. The ambiguity arises from the plastic deformation and the fracture of metal film during the delamination tests.

The fundamental requirements for delamination tests for sub-micron thin films are as follows;

Plastic deformation and the fracture of thin films do not occur during the test.

The delamination crack initiates at the edge of the interface. The

P. Ståhle and K.G. Sundin (eds.), IUTAM Symposium on Field Analyses for Determination of Material Parameters - Experimental and Numerical Aspects, 231–238.

stress intensity at the edge must be evaluated by a proper analysis.

Sufficient stress to cause delamination can be applied on the interface.

It is easy to handle the specimen.

In this study, an evaluation method, which satisfies the above requirements, is proposed, and validity is examined for the interface between films on an LSI substrate. The stress intensity factor near the free edge in the multi-layered specimen is analyzed by the boundary element method (BEM).

2. EVALUATION METHOD OF INTERFACE STRENGTH

2.1 Experimental Procedure

The material tested is shown in Fig.2 (a). The material consists of thin films, a conductor metal of copper (Cu) and a barrier layer of tantalum nitride (TaN), on a silicon substrate with an insulator of SiO2. We focus on the interface between Cu and TaN (Cu/TaN interface). The square plate of 5 mmx5 mm is cut from the silicon wafer with thin films, and a cantilever of stainless steel is carefully glued to the upper surface of the specimen, as shown in Fig.2 (b). The load, P, is applied at the edge of the cantilever. Since the steel and the silicon are relatively stiff, the deformation of the soft metal film is fully constrained by the cantilever and the substrate. This is suitable method for analysing the interface in an advanced LSI chip because the plastic deformation is restrained better for thinner films. The films surrounding the glue region are removed from the substrate before the test to avoid the affect of film fracture. The bending moment on the cantilever causes the delamination crack initiation from the edge where the stress concentrates.

The test apparatus is a remodeled micro-Vickers hardness tester as shown in Fig 3. The indenter is spherical, and the load, *P*, and the displacement at the loading point, δ, are continuously measured by means of a load cell and a linear variable differential transformer during the test, respectively. The experiments are carried out in an air at room temperature under the constant loading rate of 0.2N/s by the testing machine with an electro-magnetic loading device. Five experiments (Table 1) are conducted in order to check the scatter of evaluated strength. After the experiment, the fracture surface is observed by means of an optical microscope and a scanning electron microscopy (SEM), and the elements on the surface are examined by means of an auger electron spectroscopy (AES).

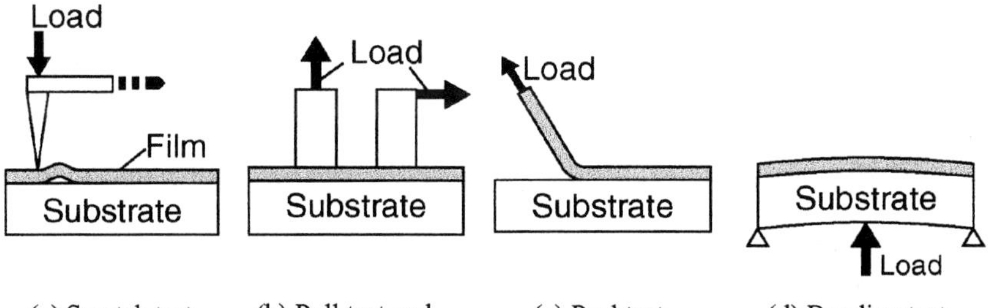

(a) Scratch test (b) Pull test and topple test (c) Peel test (d) Bending test

Figure 1. Schematic illustrations of evaluation method for interface strength between thin film and substrate.

Table 1 Experimental conditions, delamination load and toughness of interface at edge.

Test number	L_0, mm	Delamination load P_c, N	Nominal stress σ_c, MPa	α, $m^{\lambda-2}$	Toughness at the edge K_{fc}, $MPam^{-\lambda}$
1	1.4	0.9	0.33	9.5	9.1
2	1.5	1.1	0.34	8.4	8.2
3	1.6	1.0	0.33	7.9	8.9
4	1.7	1.2	0.36	7.6	9.0
5	2.1	1.5	0.37	6.6	9.9

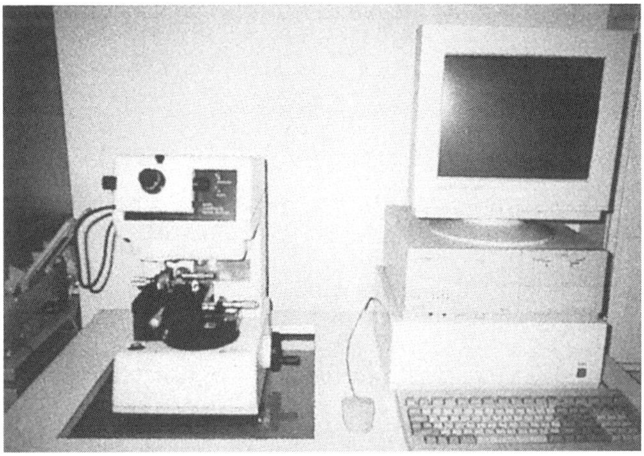

Fig. 3. Testing apparatus.

2.2 Procedure of Stress Analysis

The stress distribution along the interfaces is calculated by the boundary element method (BEM). For the configuration of specimen and the loading system, the analysis is carried out in a two-dimensional model. It is difficult to obtain the accurate stress distribution in the multi-layered thin films by usual BEM because the boundary integral includes a singular term, and this brings numerical error into the calculation of the traction and displacement along the film interface. We adopted the technique developed by Arai and Adachi [6], in which the singularity is removed by superposing a reference field equation.

The thickness of glue epoxy is set to 10μm. The real thickness in the specimen observed was between 8 and 10 μm. Preliminary analysis was carried out for the specimen with different thickness of epoxy; this revealed that the thickness had little effect on the stress distribution along the Cu/TaN interface.

Figure 4 illustrates the mesh division used for the analysis. The load, P, is applied at the node indicated by the arrow in the figure. Because the singular stress field appears near the junctions of the interface and the surface due to the free edge effect, the region is carefully divided into a fine mesh. Especially, the length of the element near the edge of Cu/TaN interface is set to 5nm in order to reproduce the stress field precisely. The elastic constants used in the calculation are listed in Table 2.

Table 2 Elastic constants.

Material	E, Gpa	ν
Cu	130	0.34
TaN	350	0.35
SiO_2	92	0.30
Si	170	0.30
Epoxy	2.5	0.30
Stainless steal	200	0.30

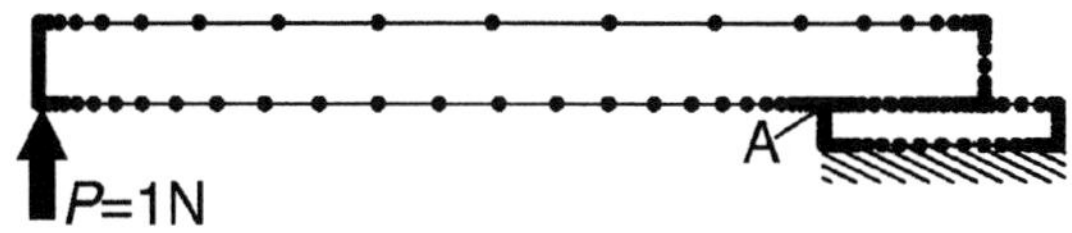

(a) Mesh division and boundary conditions

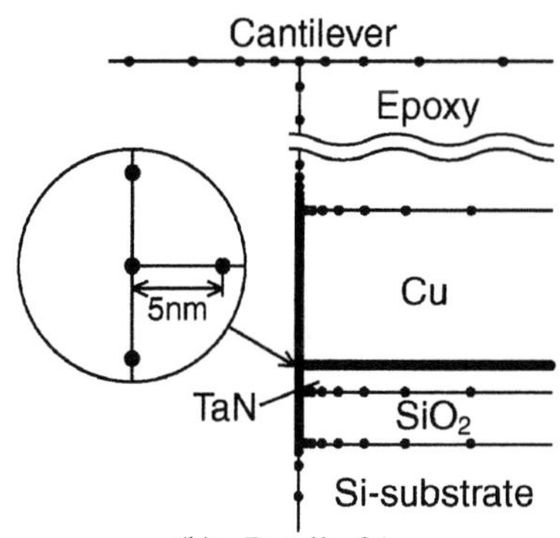

(b) Detail of A

Figure 4. Mesh division for BEM analysis.

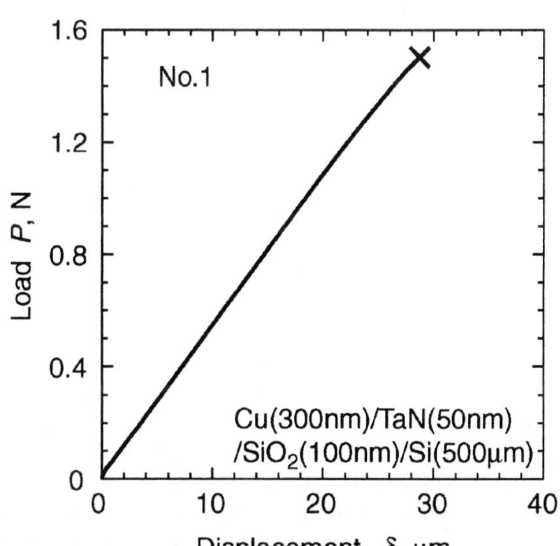

Figure 5. Relationship between load and displacement at load point.

3 RESULTS AND DISCUSSION

Figure 5 shows the P - δ curve during the delamination test. The linear relation suggests that little plastic deformation takes place in the sandwiched specimen. Table 1 list the delamination load, P_c, and the nominal fracture stress, $\sigma_c = P_c / A$, where A is the glued area. The small scatter in the strength implies the excellent repeatability of this method.

Figure 6 shows the fractographs of the specimen. Both of the fracture surfaces are smooth and flat. The magnified photographs taken by the SEM indicates that there is no trace of crystallographic slips or fracture in the Cu and TaN films. Thus, the stiffness of the cantilever effectively prevents the plastic deformation of the films during the delamination. The gloss on the surface of the cantilever side suggests the copper film while the green layer covers the surface of substrate side. The AES shown in Fig. 7 confirms that the fracture surfaces consist of Cu and TaN, respectively. Thus, the delamination crack propagates along the Cu/TaN interface.

Figure 8 (a) plots the distribution of the normal stress, σ_{yy}, along the Cu/TaN interface at the delamination. The tensile stress concentrates at the left edges of the interface, while the compressive stress dominates the other edge. The tensile stress near the left edge stems from the mismatch in elastic coefficients of the materials, and the singular stress field is formulated by [7].

$$\sigma_{ij} = K_f \frac{1}{r^{\lambda}} , \tag{1}$$

where K_f is the stress intensity factor depending on the applied load and the size of specimen, r is the distance from the left free edge, λ is parameter depending on the materials and the shape of the junction of surface and interface. On the basis of Bogy's analysis [7], the magnitude of λ must be 0.08 for the junction of the specimen. As shown in Fig.8 (b), this singularity is successfully reproduced by the BEM calculation.

Moreover, the figure shows that the singular field governs only the region, r<40nm, which is comparable with the thickness of the TaN film, and is much smaller than that of the Cu one.

As an analogy to the evaluation of toughness for a cracked body on the basis of the fracture mechanics concept, the criterion of interface crack initiation (the interface toughness at the free edge) might be evaluated by the stress singularity near the free edge. This means that the toughness is given by the critical magnitude of K_f at the delamination. Using the result of BEM analysis, we correlated K_f for the specimen with the applied load:

$$K_f = \alpha P , \tag{2}$$

where α has the dimension of $m^{\lambda-2}$, and the magnitude listed in Table 1. The critical stress intensity factor, K_{fc}, can be calculated by substituting the delamination load into Eq.(2). The toughness at the free edge of Cu/TaN interface by the experiments (Table1) is about 9.0MPam$^{-\lambda}$ (λ=0.08).

To verify the validity of the fracture mechanics concept, it is necessary to examine the interface strength by the tests using another configuration of the

specimen,

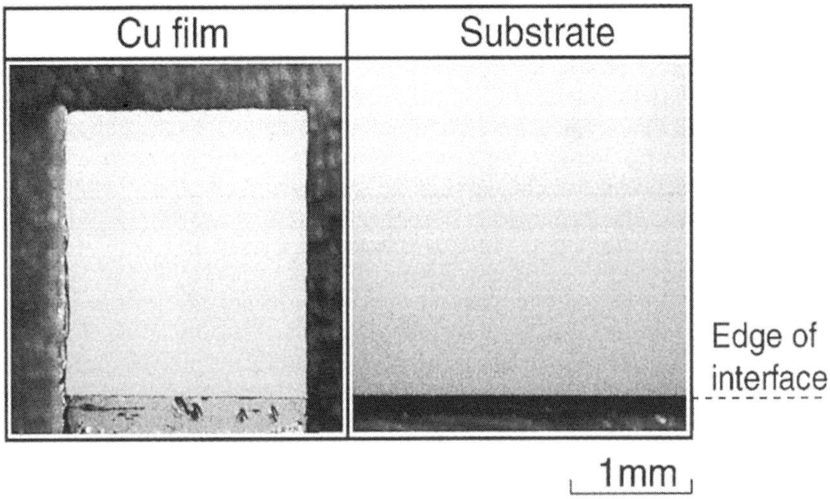

(a) Fracture surfaces

(b) Magnified view near the delamination edge

Figure 6. Fractographs of specimen

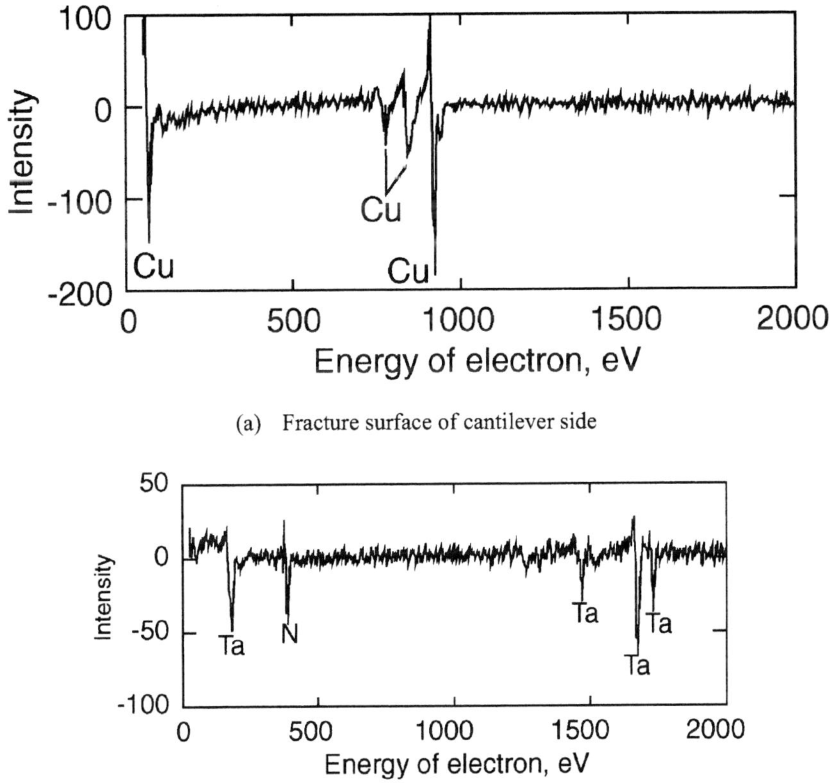

(a) Fracture surface of cantilever side

(b) Fracture surface of substrate side

Figure 7. Auger electron spectrums of the fracture surfaces.

which possesses different global stress distribution but the same stress singularity (the same magnitude of λ) near the free edge. We conducted such tests for several shapes of specimen and the results are discussed in elsewhere [8]. These experiments show that the stress intensity factor, which governs the stress field in only 40nm from the edge, controls the delamination crack initiation.

4. CONCLUSIONS

An evaluation method of interface strength between thin films is proposed. The plastic deformation and the fracture of films are carefully avoided by the sandwiched structure of the specimen during the delamination test.

The delamination crack initiation from the edge of interface is governed by the stress intensity factor, K_f, which represents the stress singular field due to the free edge effect.

The critical toughness of Cu/TaN interface is successfully evaluated by the proposed method and the BEM analysis.

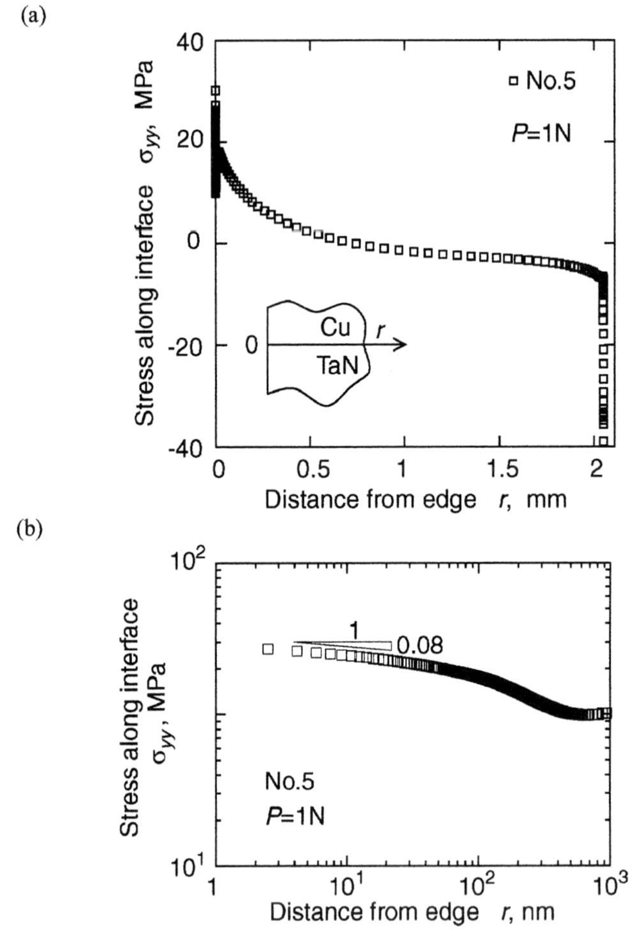

Figure 8. Stress near the edge of interface Distributions of the normal stress along the Cu and TaN interface.

References

1. A. Kinbara, E. Kusano, T. Kamiya, I. Kondo and O. Takenaka, *Thin Solid Films*, Vol. 317, pp. 165-168, 1998.
2. M. T. Laugier, *Thin Solid Films*, Vol. 117, pp.243-249, 1984.
3. J. Salem and F. Seqeda, *J. Vac. Sci. Technol.*, Vol. 18, No.2, pp149-152, 1981.
4. A. J. Kinloch, C. C. Lau and J. G. Williams, *Int. J. Fracture*, Vol.66, pp.45-70, 1994.
5. Q. Ma, H. Fujimoto, P. Flinn, V. Jain, F. Adibi-Rizi, F. Moghadam and R. H. Dauskardt, *Materials Reliability in Microelectronics V* (*MRS Symp. Proc.*), Vol. 391, pp. 91-96, 1995.
6. M. Arai, T. Adachi, K. Nakayama and H. Matsumoto, *Trans. JSME*, Vol. 62, No. 597, pp. 1212-1218, 1996.
7. D. B. Bogy, *J. Appl. Mech.*, Vol.35, pp.146-154, 1968.
8. T. Kitamura and T. Shibutani, In preparation.

AUTHORS

Mechanics

SOLID MECHANICS AND ITS APPLICATIONS

Series Editor: G.M.L. Gladwell

Aims and Scope of the Series

The fundamental questions arising in mechanics are: *Why?*, *How?*, and *How much?* The aim of this series is to provide lucid accounts written by authoritative researchers giving vision and insight in answering these questions on the subject of mechanics as it relates to solids. The scope of the series covers the entire spectrum of solid mechanics. Thus it includes the foundation of mechanics; variational formulations; computational mechanics; statics, kinematics and dynamics of rigid and elastic bodies; vibrations of solids and structures; dynamical systems and chaos; the theories of elasticity, plasticity and viscoelasticity; composite materials; rods, beams, shells and membranes; structural control and stability; soils, rocks and geomechanics; fracture; tribology; experimental mechanics; biomechanics and machine design.

1. R.T. Haftka, Z. Gürdal and M.P. Kamat: *Elements of Structural Optimization*. 2nd rev.ed., 1990 ISBN 0-7923-0608-2
2. J.J. Kalker: *Three-Dimensional Elastic Bodies in Rolling Contact*. 1990 ISBN 0-7923-0712-7
3. P. Karasudhi: *Foundations of Solid Mechanics*. 1991 ISBN 0-7923-0772-0
4. *Not published*
5. *Not published.*
6. J.F. Doyle: *Static and Dynamic Analysis of Structures*. With an Emphasis on Mechanics and Computer Matrix Methods. 1991 ISBN 0-7923-1124-8; Pb 0-7923-1208-2
7. O.O. Ochoa and J.N. Reddy: *Finite Element Analysis of Composite Laminates*. ISBN 0-7923-1125-6
8. M.H. Aliabadi and D.P. Rooke: *Numerical Fracture Mechanics*. ISBN 0-7923-1175-2
9. J. Angeles and C.S. López-Cajún: *Optimization of Cam Mechanisms*. 1991 ISBN 0-7923-1355-0
10. D.E. Grierson, A. Franchi and P. Riva (eds.): *Progress in Structural Engineering*. 1991 ISBN 0-7923-1396-8
11. R.T. Haftka and Z. Gürdal: *Elements of Structural Optimization*. 3rd rev. and exp. ed. 1992 ISBN 0-7923-1504-9; Pb 0-7923-1505-7
12. J.R. Barber: *Elasticity*. 1992 ISBN 0-7923-1609-6; Pb 0-7923-1610-X
13. H.S. Tzou and G.L. Anderson (eds.): *Intelligent Structural Systems*. 1992 ISBN 0-7923-1920-6
14. E.E. Gdoutos: *Fracture Mechanics*. An Introduction. 1993 ISBN 0-7923-1932-X
15. J.P. Ward: *Solid Mechanics*. An Introduction. 1992 ISBN 0-7923-1949-4
16. M. Farshad: *Design and Analysis of Shell Structures*. 1992 ISBN 0-7923-1950-8
17. H.S. Tzou and T. Fukuda (eds.): *Precision Sensors, Actuators and Systems*. 1992 ISBN 0-7923-2015-8
18. J.R. Vinson: *The Behavior of Shells Composed of Isotropic and Composite Materials*. 1993 ISBN 0-7923-2113-8
19. H.S. Tzou: *Piezoelectric Shells*. Distributed Sensing and Control of Continua. 1993 ISBN 0-7923-2186-3
20. W. Schiehlen (ed.): *Advanced Multibody System Dynamics*. Simulation and Software Tools. 1993 ISBN 0-7923-2192-8
21. C.-W. Lee: *Vibration Analysis of Rotors*. 1993 ISBN 0-7923-2300-9
22. D.R. Smith: *An Introduction to Continuum Mechanics*. 1993 ISBN 0-7923-2454-4
23. G.M.L. Gladwell: *Inverse Problems in Scattering*. An Introduction. 1993 ISBN 0-7923-2478-1

Mechanics

SOLID MECHANICS AND ITS APPLICATIONS

Series Editor: G.M.L. Gladwell

24. G. Prathap: *The Finite Element Method in Structural Mechanics*. 1993 ISBN 0-7923-2492-7
25. J. Herskovits (ed.): *Advances in Structural Optimization*. 1995 ISBN 0-7923-2510-9
26. M.A. González-Palacios and J. Angeles: *Cam Synthesis*. 1993 ISBN 0-7923-2536-2
27. W.S. Hall: *The Boundary Element Method*. 1993 ISBN 0-7923-2580-X
28. J. Angeles, G. Hommel and P. Kovács (eds.): *Computational Kinematics*. 1993 ISBN 0-7923-2585-0
29. A. Curnier: *Computational Methods in Solid Mechanics*. 1994 ISBN 0-7923-2761-6
30. D.A. Hills and D. Nowell: *Mechanics of Fretting Fatigue*. 1994 ISBN 0-7923-2866-3
31. B. Tabarrok and F.P.J. Rimrott: *Variational Methods and Complementary Formulations in Dynamics*. 1994 ISBN 0-7923-2923-6
32. E.H. Dowell (ed.), E.F. Crawley, H.C. Curtiss Jr., D.A. Peters, R. H. Scanlan and F. Sisto: *A Modern Course in Aeroelasticity*. Third Revised and Enlarged Edition. 1995 ISBN 0-7923-2788-8; Pb: 0-7923-2789-6
33. A. Preumont: *Random Vibration and Spectral Analysis*. 1994 ISBN 0-7923-3036-6
34. J.N. Reddy (ed.): *Mechanics of Composite Materials*. Selected works of Nicholas J. Pagano. 1994 ISBN 0-7923-3041-2
35. A.P.S. Selvadurai (ed.): *Mechanics of Poroelastic Media*. 1996 ISBN 0-7923-3329-2
36. Z. Mróz, D. Weichert, S. Dorosz (eds.): *Inelastic Behaviour of Structures under Variable Loads*. 1995 ISBN 0-7923-3397-7
37. R. Pyrz (ed.): *IUTAM Symposium on Microstructure-Property Interactions in Composite Materials*. Proceedings of the IUTAM Symposium held in Aalborg, Denmark. 1995 ISBN 0-7923-3427-2
38. M.I. Friswell and J.E. Mottershead: *Finite Element Model Updating in Structural Dynamics*. 1995 ISBN 0-7923-3431-0
39. D.F. Parker and A.H. England (eds.): *IUTAM Symposium on Anisotropy, Inhomogeneity and Nonlinearity in Solid Mechanics*. Proceedings of the IUTAM Symposium held in Nottingham, U.K. 1995 ISBN 0-7923-3594-5
40. J.-P. Merlet and B. Ravani (eds.): *Computational Kinematics '95*. 1995 ISBN 0-7923-3673-9
41. L.P. Lebedev, I.I. Vorovich and G.M.L. Gladwell: *Functional Analysis*. Applications in Mechanics and Inverse Problems. 1996 ISBN 0-7923-3849-9
42. J. Menčik: *Mechanics of Components with Treated or Coated Surfaces*. 1996 ISBN 0-7923-3700-X
43. D. Bestle and W. Schiehlen (eds.): *IUTAM Symposium on Optimization of Mechanical Systems*. Proceedings of the IUTAM Symposium held in Stuttgart, Germany. 1996 ISBN 0-7923-3830-8
44. D.A. Hills, P.A. Kelly, D.N. Dai and A.M. Korsunsky: *Solution of Crack Problems*. The Distributed Dislocation Technique. 1996 ISBN 0-7923-3848-0
45. V.A. Squire, R.J. Hosking, A.D. Kerr and P.J. Langhorne: *Moving Loads on Ice Plates*. 1996 ISBN 0-7923-3953-3
46. A. Pineau and A. Zaoui (eds.): *IUTAM Symposium on Micromechanics of Plasticity and Damage of Multiphase Materials*. Proceedings of the IUTAM Symposium held in Sèvres, Paris, France. 1996 ISBN 0-7923-4188-0
47. A. Naess and S. Krenk (eds.): *IUTAM Symposium on Advances in Nonlinear Stochastic Mechanics*. Proceedings of the IUTAM Symposium held in Trondheim, Norway. 1996 ISBN 0-7923-4193-7
48. D. Ieşan and A. Scalia: *Thermoelastic Deformations*. 1996 ISBN 0-7923-4230-5

Mechanics

SOLID MECHANICS AND ITS APPLICATIONS

Series Editor: G.M.L. Gladwell

49. J.R. Willis (ed.): *IUTAM Symposium on Nonlinear Analysis of Fracture*. Proceedings of the IUTAM Symposium held in Cambridge, U.K. 1997 ISBN 0-7923-4378-6
50. A. Preumont: *Vibration Control of Active Structures*. An Introduction. 1997 ISBN 0-7923-4392-1
51. G.P. Cherepanov: *Methods of Fracture Mechanics: Solid Matter Physics*. 1997 ISBN 0-7923-4408-1
52. D.H. van Campen (ed.): *IUTAM Symposium on Interaction between Dynamics and Control in Advanced Mechanical Systems*. Proceedings of the IUTAM Symposium held in Eindhoven, The Netherlands. 1997 ISBN 0-7923-4429-4
53. N.A. Fleck and A.C.F. Cocks (eds.): *IUTAM Symposium on Mechanics of Granular and Porous Materials*. Proceedings of the IUTAM Symposium held in Cambridge, U.K. 1997 ISBN 0-7923-4553-3
54. J. Roorda and N.K. Srivastava (eds.): *Trends in Structural Mechanics*. Theory, Practice, Education. 1997 ISBN 0-7923-4603-3
55. Yu.A. Mitropolskii and N. Van Dao: *Applied Asymptotic Methods in Nonlinear Oscillations*. 1997 ISBN 0-7923-4605-X
56. C. Guedes Soares (ed.): *Probabilistic Methods for Structural Design*. 1997 ISBN 0-7923-4670-X
57. D. François, A. Pineau and A. Zaoui: *Mechanical Behaviour of Materials*. Volume I: Elasticity and Plasticity. 1998 ISBN 0-7923-4894-X
58. D. François, A. Pineau and A. Zaoui: *Mechanical Behaviour of Materials*. Volume II: Viscoplasticity, Damage, Fracture and Contact Mechanics. 1998 ISBN 0-7923-4895-8
59. L.T. Tenek and J. Argyris: *Finite Element Analysis for Composite Structures*. 1998 ISBN 0-7923-4899-0
60. Y.A. Bahei-El-Din and G.J. Dvorak (eds.): *IUTAM Symposium on Transformation Problems in Composite and Active Materials*. Proceedings of the IUTAM Symposium held in Cairo, Egypt. 1998 ISBN 0-7923-5122-3
61. I.G. Goryacheva: *Contact Mechanics in Tribology*. 1998 ISBN 0-7923-5257-2
62. O.T. Bruhns and E. Stein (eds.): *IUTAM Symposium on Micro- and Macrostructural Aspects of Thermoplasticity*. Proceedings of the IUTAM Symposium held in Bochum, Germany. 1999 ISBN 0-7923-5265-3
63. F.C. Moon: *IUTAM Symposium on New Applications of Nonlinear and Chaotic Dynamics in Mechanics*. Proceedings of the IUTAM Symposium held in Ithaca, NY, USA. 1998 ISBN 0-7923-5276-9
64. R. Wang: *IUTAM Symposium on Rheology of Bodies with Defects*. Proceedings of the IUTAM Symposium held in Beijing, China. 1999 ISBN 0-7923-5297-1
65. Yu.I. Dimitrienko: *Thermomechanics of Composites under High Temperatures*. 1999 ISBN 0-7923-4899-0
66. P. Argoul, M. Frémond and Q.S. Nguyen (eds.): *IUTAM Symposium on Variations of Domains and Free-Boundary Problems in Solid Mechanics*. Proceedings of the IUTAM Symposium held in Paris, France. 1999 ISBN 0-7923-5450-8
67. F.J. Fahy and W.G. Price (eds.): *IUTAM Symposium on Statistical Energy Analysis*. Proceedings of the IUTAM Symposium held in Southampton, U.K. 1999 ISBN 0-7923-5457-5
68. H.A. Mang and F.G. Rammerstorfer (eds.): *IUTAM Symposium on Discretization Methods in Structural Mechanics*. Proceedings of the IUTAM Symposium held in Vienna, Austria. 1999 ISBN 0-7923-5591-1

Mechanics

SOLID MECHANICS AND ITS APPLICATIONS

Series Editor: G.M.L. Gladwell

69. P. Pedersen and M.P. Bendsøe (eds.): *IUTAM Symposium on Synthesis in Bio Solid Mechanics*. Proceedings of the IUTAM Symposium held in Copenhagen, Denmark. 1999 ISBN 0-7923-5615-2
70. S.K. Agrawal and B.C. Fabien: *Optimization of Dynamic Systems*. 1999 ISBN 0-7923-5681-0
71. A. Carpinteri: *Nonlinear Crack Models for Nonmetallic Materials*. 1999 ISBN 0-7923-5750-7
72. F. Pfeifer (ed.): *IUTAM Symposium on Unilateral Multibody Contacts*. Proceedings of the IUTAM Symposium held in Munich, Germany. 1999 ISBN 0-7923-6030-3
73. E. Lavendelis and M. Zakrzhevsky (eds.): *IUTAM/IFToMM Symposium on Synthesis of Nonlinear Dynamical Systems*. Proceedings of the IUTAM/IFToMM Symposium held in Riga, Latvia. 2000 ISBN 0-7923-6106-7
74. J.-P. Merlet: *Parallel Robots*. 2000 ISBN 0-7923-6308-6
75. J.T. Pindera: *Techniques of Tomographic Isodyne Stress Analysis*. 2000 ISBN 0-7923-6388-4
76. G.A. Maugin, R. Drouot and F. Sidoroff (eds.): *Continuum Thermomechanics*. The Art and Science of Modelling Material Behaviour. 2000 ISBN 0-7923-6407-4
77. N. Van Dao and E.J. Kreuzer (eds.): *IUTAM Symposium on Recent Developments in Non-linear Oscillations of Mechanical Systems*. 2000 ISBN 0-7923-6470-8
78. S.D. Akbarov and A.N. Guz: *Mechanics of Curved Composites*. 2000 ISBN 0-7923-6477-5
79. M.B. Rubin: *Cosserat Theories: Shells, Rods and Points*. 2000 ISBN 0-7923-6489-9
80. S. Pellegrino and S.D. Guest (eds.): *IUTAM-IASS Symposium on Deployable Structures: Theory and Applications*. Proceedings of the IUTAM-IASS Symposium held in Cambridge, U.K., 6–9 September 1998. 2000 ISBN 0-7923-6516-X
81. A.D. Rosato and D.L. Blackmore (eds.): *IUTAM Symposium on Segregation in Granular Flows*. Proceedings of the IUTAM Symposium held in Cape May, NJ, U.S.A., June 5–10, 1999. 2000 ISBN 0-7923-6547-X
82. A. Lagarde (ed.): *IUTAM Symposium on Advanced Optical Methods and Applications in Solid Mechanics*. Proceedings of the IUTAM Symposium held in Futuroscope, Poitiers, France, August 31–September 4, 1998. 2000 ISBN 0-7923-6604-2
83. D. Weichert and G. Maier (eds.): *Inelastic Analysis of Structures under Variable Loads*. Theory and Engineering Applications. 2000 ISBN 0-7923-6645-X
84. T.-J. Chuang and J.W. Rudnicki (eds.): *Multiscale Deformation and Fracture in Materials and Structures*. The James R. Rice 60th Anniversary Volume. 2001 ISBN 0-7923-6718-9
85. S. Narayanan and R.N. Iyengar (eds.): *IUTAM Symposium on Nonlinearity and Stochastic Structural Dynamics*. Proceedings of the IUTAM Symposium held in Madras, Chennai, India, 4–8 January 1999 ISBN 0-7923-6733-2
86. S. Murakami and N. Ohno (eds.): *IUTAM Symposium on Creep in Structures*. Proceedings of the IUTAM Symposium held in Nagoya, Japan, 3-7 April 2000. 2001 ISBN 0-7923-6737-5
87. W. Ehlers (ed.): *IUTAM Symposium on Theoretical and Numerical Methods in Continuum Mechanics of Porous Materials*. Proceedings of the IUTAM Symposium held at the University of Stuttgart, Germany, September 5-10, 1999. 2001 ISBN 0-7923-6766-9
88. D. Durban, D. Givoli and J.G. Simmonds (eds.): *Advances in the Mechanis of Plates and Shells The Avinoam Libai Anniversary Volume*. 2001 ISBN 0-7923-6785-5
89. U. Gabbert and H.-S. Tzou (eds.): *IUTAM Symposium on Smart Structures and Structonic Systems*. Proceedings of the IUTAM Symposium held in Magdeburg, Germany, 26–29 September 2000. 2001 ISBN 0-7923-6968-8

Mechanics

SOLID MECHANICS AND ITS APPLICATIONS

Series Editor: G.M.L. Gladwell

90. Y. Ivanov, V. Cheshkov and M. Natova: *Polymer Composite Materials – Interface Phenomena & Processes*. 2001 ISBN 0-7923-7008-2
91. R.C. McPhedran, L.C. Botten and N.A. Nicorovici (eds.): *IUTAM Symposium on Mechanical and Electromagnetic Waves in Structured Media*. Proceedings of the IUTAM Symposium held in Sydney, NSW, Australia, 18-22 Januari 1999. 2001 ISBN 0-7923-7038-4
92. D.A. Sotiropoulos (ed.): *IUTAM Symposium on Mechanical Waves for Composite Structures Characterization*. Proceedings of the IUTAM Symposium held in Chania, Crete, Greece, June 14-17, 2000. 2001 ISBN 0-7923-7164-X
93. V.M. Alexandrov and D.A. Pozharskii: *Three-Dimensional Contact Problems*. 2001 ISBN 0-7923-7165-8
94. J.P. Dempsey and H.H. Shen (eds.): *IUTAM Symposium on Scaling Laws in Ice Mechanics and Ice Dynamics*. Proceedings of the IUTAM Symposium held in Fairbanks, Alaska, U.S.A., 13-16 June 2000. 2001 ISBN 1-4020-0171-1
95. U. Kirsch: *Design-Oriented Analysis of Structures*. A Unified Approach. 2002 ISBN 1-4020-0443-5
96. A. Preumont: *Vibration Control of Active Structures*. An Introduction (2nd Edition). 2002 ISBN 1-4020-0496-6
97. B.L. Karihaloo (ed.): *IUTAM Symposium on Analytical and Computational Fracture Mechanics of Non-Homogeneous Materials*. Proceedings of the IUTAM Symposium held in Cardiff, U.K., 18-22 June 2001. 2002 ISBN 1-4020-0510-5
98. S.M. Han and H. Benaroya: *Nonlinear and Stochastic Dynamics of Compliant Offshore Structures*. 2002 ISBN 1-4020-0573-3
99. A.M. Linkov: *Boundary Integral Equations in Elasticity Theory*. 2002 ISBN 1-4020-0574-1
100. L.P. Lebedev, I.I. Vorovich and G.M.L. Gladwell: *Functional Analysis*. Applications in Mechanics and Inverse Problems (2nd Edition). 2002 ISBN 1-4020-0667-5; Pb: 1-4020-0756-6
101. Q.P. Sun (ed.): *IUTAM Symposium on Mechanics of Martensitic Phase Transformation in Solids*. Proceedings of the IUTAM Symposium held in Hong Kong, China, 11-15 June 2001. 2002 ISBN 1-4020-0741-8
102. M.L. Munjal (ed.): *IUTAM Symposium on Designing for Quietness*. Proceedings of the IUTAM Symposium held in Bangkok, India, 12-14 December 2000. 2002 ISBN 1-4020-0765-5
103. J.A.C. Martins and M.D.P. Monteiro Marques (eds.): *Contact Mechanics*. Proceedings of the 3rd Contact Mechanics International Symposium, Praia da Consolação, Peniche, Portugal, 17-21 June 2001. 2002 ISBN 1-4020-0811-2
104. H.R. Drew and S. Pellegrino (eds.): *New Approaches to Structural Mechanics, Shells and Biological Structures*. 2002 ISBN 1-4020-0862-7
105. J.R. Vinson and R.L. Sierakowski: *The Behavior of Structures Composed of Composite Materials*. Second Edition. 2002 ISBN 1-4020-0904-6
106. Not yet published.
107. J.R. Barber: *Elasticity*. Second Edition. 2002 ISBN Hb 1-4020-0964-X; Pb 1-4020-0966-6
108. C. Miehe (ed.): *IUTAM Symposium on Computational Mechanics of Solid Materials at Large Strains*. Proceedings of the IUTAM Symposium held in Stuttgart, Germany, 20-24 August 2001. 2003 ISBN 1-4020-1170-9

Mechanics

SOLID MECHANICS AND ITS APPLICATIONS

Series Editor: G.M.L. Gladwell

109. P. Ståhle and K.G. Sundin (eds.): *IUTAM Symposium on Field Analyses for Determination of Material Parameters – Experimental and Numerical Aspects.* Proceedings of the IUTAM Symposium held in Abisko National Park, Kiruna, Sweden, July 31 – August 4, 2000. 2003
ISBN 1-4020-1283-7

Kluwer Academic Publishers – Dordrecht / Boston / London